Fundamentals of Low-Dimensional Carbon Nanomaterials

MATERIALS RESEARCH SOCIETY
SYMPOSIUM PROCEEDINGS VOLUME 1284

Fundamentals of Low-Dimensional Carbon Nanomaterials

Symposium held November 29–December 3, Boston, Massachusetts, U.S.A.

EDITORS

John J. Boeckl
Air Force Research Laboratory
Wright-Patterson AFB, Ohio, U.S.A.

Mark Rümmeli
Leibniz Institute, IFW Dresden
Dresden, Germany

Weijie Lu
Fisk University
Nashville, Tennessee, U.S.A.

Jamie Warner
University of Oxford
Oxford, United Kingdom

Materials Research Society
Warrendale, Pennsylvania

CAMBRIDGE
UNIVERSITY PRESS

CAMBRIDGE UNIVERSITY PRESS
Cambridge, New York, Melbourne, Madrid, Cape Town,
Singapore, São Paulo, Delhi, Mexico City

Cambridge University Press
32 Avenue of the Americas, New York NY 10013-2473, USA

Published in the United States of America by Cambridge University Press, New York

www.cambridge.org
Information on this title: www.cambridge.org/9781107406667

Materials Research Society
506 Keystone Drive, Warrendale, PA 15086
http://www.mrs.org

© Materials Research Society 2011

First published 2011
First paperback edition 2012

Single article reprints from this publication are available through
University Microfilms Inc., 300 North Zeeb Road, Ann Arbor, MI 48106

CODEN: MRSPDH

ISBN 978-1-605-11261-9 Hardback
ISBN 978-1-107-40666-7 Paperback

CONTENTS

*Invited Paper

NOVEL STRUCTURES AND PROPERTIES OF LOW DIMENSIONAL CARBON NANOSTRUCTURES

PREFACE

Symposium C, "Fundamentals of Low-Dimensional Carbon Nanomaterials," was held Nov. 29–Dec. 3 at the 2010 MRS Fall Meeting in Boston, Massachusetts. This resultant proceedings volume comprises 27 manuscripts with topics including growth techniques for CNTs and graphene, structural characterization, novel properties, and interface & surface structures. This was the first symposium at the MRS meeting which was devoted solely to fundamental issues of low-dimensional carbon nanomaterials. Device applications of carbon nanostructures were excluded from this symposium.

Low-dimensional carbon nanostructures exhibit a rich structural diversity from zero-dimensional C60, one-dimensional carbon nanotubes (CNTs), and two-dimensional graphene and graphite oxides. These low-dimensional carbon nanostructures are at the forefront of materials science and provide a platform for understanding the growth mechanisms and properties of nanostructures in general. They exhibit novel properties with endless potential applications from high-speed electronics to high-performance composites.

Although low-dimensional carbon nanomaterials have attracted great interest in the research community, the applications and commercialization of graphene and CNTs have, to date, not been as successful as anticipated. The need for significant improvements in material quality and structural uniformity exists. Other areas that need further understanding include the atomic scale growth mechanisms, structural control of various graphitic nanostructures, the chemistry of graphitic hexagonal structures, and graphitization engineering in low dimensions. Without comprehending the basic growth mechanisms and techniques to control atomic structure, the promise of future applications will be difficult to achieve.

The editors would like to thank the authors of the manuscripts. MRS meetings have become one of the most important forums for carbon nanomaterials. The challenges in fundamental issues of low-dimensional carbon nanomaterials have a great impact not only on carbon material science but also on the general fields of nanoscience and nanoengineering. This volume is a useful resource to share interests within this broad research community.

John J. Boeckl
Mark Rümmeli
Weijie Lu
Jamie Warner

February 2011

MATERIALS RESEARCH SOCIETY SYMPOSIUM PROCEEDINGS

MATERIALS RESEARCH SOCIETY SYMPOSIUM PROCEEDINGS

Prior Materials Research Society Symposium Proceedings available by contacting Materials Research Society

Growth of Graphene/Substrate Structures

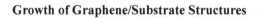

Mater. Res. Soc. Symp. Proc. Vol. 1284 © 2011 Materials Research Society
DOI: 10.1557/opl.2011.638

Graphene growth on SiC and other substrates using carbon sources

W. C. Mitchel,[1] J. H. Park,[1] Howard E. Smith,[1] L. Grazulis,[1] S. Mou, D. Tomich[1], K. Eyink[1] and Said Elhamri[2]

[1]Air Force Research Laboratory, Materials and Manufacturing Directorate, WPAFB, Dayton, OH 45433-7707, USA

2Department of Physics, University of Dayton, Dayton, OH 45469, USA

ABSTRACT

Direct deposition of graphene from carbon sources on foreign substrates without the use of metal catalysts is shown to be an effective process with several advantages over other growth techniques. Carbon source molecular beam epitaxy (CMBE) in particular provides an additional control parameter in carbon flux and enables growth on substrates other than SiC, including oxidized Si and sapphire. CMBE using thermally evaporated C_{60} and a heated graphite filament on SiC is reported here. The graphene films were characterized by Raman spectroscopy, X-ray photoelectron spectroscopy, atomic force microscopy and Hall effect. Graphene films on Si-face SiC grown using the C_{60} source have Bernal-like stacking and n-type conduction while those grown using the graphite filament have turbostratic stacking and p-type conduction. The sheet concentration for both n- and p-type doping is linearly dependent on film thickness.

INTRODUCTION

Since the isolation of single layer graphene was reported by Novoselov et al. in 2004 [1] a variety of techniques have been used to grow or fabricated this material. Novoselov et al. used mechanical exfoliation from natural or artificial graphite. The technique for removing a few layers from the top of a graphite sample with tape was well known in the scanning tunneling microscopy community for some time but the graphene was always discarded with the tape. Mechanical exfoliation still produces the highest electrical quality graphene but the largest areas exfoliated to date are about 100 μm x 100 μm and no one has yet developed a reliable process for precisely depositing graphene at specific locations on the substrate by this approach. Perhaps the oldest approach for producing graphene is chemical vapor deposition on metal films. Karu and Beer [2] reported growth of "crystalline films of graphite by pyrolysis of methane on hot single crystal nickel" in 1966. However, this technique did not become a viable means of producing electronic grade graphene until the work of Kim et al. [3] and Li et al. [4]. Growth on both Ni [3] and Cu [4] foils has produced high quality graphene. However, the technique requires somewhat elaborate processes to transfer the graphene from the conductive metal foils to more useful substrates such as oxidized Si and damage can result. Chemical exfoliation of graphene oxide from graphite and subsequent deposition and reduction back to graphene has been reported [5]. This technique is inexpensive and useful for applications that do not require high mobility, such as transparent conductors and interconnects. It was long known in the SiC community that annealing SiC at high temperatures resulted in the sublimation of Si and the formation of a graphite-like carbon layer on the SiC surface [6], but it took Berger et al. [7] to use the decomposition of SiC in ultra high vacuum (UHV) to produce a useful form of graphene.

Emtsev et al. [8] later demonstrated high quality growth of graphene by decomposition of SiC in an atmospheric pressure argon ambient but at the cost of higher annealing temperatures.

Recently several groups have investigated the direct deposition of graphene on foreign substrates without the use of catalysts in UHV using a variety of carbon sources. Hackley et al. [9] reported growth of graphitic films directly on Si(111) in a molecular beam epitaxy (MBE) chamber. Their carbon source was an electron beam evaporated graphite target. They reported X-ray photoelectron spectroscopy (XPS) and Raman spectroscopy measurements but not resistivity or Hall affect measurements. The disorder based Raman D band was more intense than the G band and the Raman spectra did not go out far enough to show the 2D band. Al-Temimy et al. [10] used a commercial resistively heated graphite filament as their carbon source. They used low energy electron diffraction to demonstrate graphene like surface reconstructions as well as angle resolved ultraviolet photoelectron spectroscopy and atomic force microscopy (AFM) to demonstrate the presence of graphene. Moreau et al. [11] used the same carbon source as Al-Temimy to grow graphene on SiC. Both Al-Temimy et al. and Moreau et al. first prepared the SiC surface as for UHV SiC decomposition but then turned on their carbon source rather than continuing heating the substrate for Si sublimation. Neither reported Raman spectroscopy or electrical measurements. Hwang et al. [12] reported graphene growth directly on SiC and sapphire substrates by chemical vapor deposition growth using propane at temperatures from 1350 to 1650°C. Raman measurements showed strong G and 2D bands and a weak D band for both substrates. Synchrotron X-ray measurements indicated that the stacking sequence in multilayer graphene depended on the substrate. Also using CVD but with acetylene Usachov et al. recently reported direct growth of graphene on BN films [13]. The present authors [14] reported growth in a UHV MBE chamber using both thermally evaporated C_{60} and a resistively heated graphite filament carbon sources. We report here further experiments on carbon source MBE (CMBE) of graphene using these two sources including resistivity and Hall affect measurements in addition to AFM, XPS and Raman spectroscopy measurements. We demonstrate that both the graphene layer stacking for multilayer films and the carrier type depend on the carbon source used. We also demonstrate graphene growth on SiO_2 on Si using the C_{60} source.

EXPERIMENTAL DETAILS

The growth process has been described elsewhere [14, 15]. All growths were on Si-face semi-insulating 4H SiC. Two carbon sources were used, thermally evaporated C_{60} and a resistively heated graphite filament. The C_{60} powder is heated in a conventional shuttered MBE cell to around 500°C. This is well below the decomposition temperature of around 800°C [16]. Kolodney, Tsipinyuk and Budrevich [17] report that C_{60} decomposes as $C_{60}^+ \rightarrow C_{58}^+ + C_2$ with an activation energy of about 4 eV. C_{60} has been used as carbon source in MBE growth of SiC on Si [18], SiC on SiC [19] and SiC on sapphire [20]. The graphite filament in these experiments serves the dual purpose of both substrate heater and carbon source. The back side of the substrate holder was cut open to enable higher growth temperature but also exposed the back side of the substrate to the carbon flux emitted from the heater. The heater was used as a carbon source by mounting the substrate with the surface of interest facing the heater. Resistively heated graphite has been used for some time as carbon sources for p-type doping of III-V semiconductors in MBE experiments [21]. The flux from such filaments consists mainly of C_3 molecules with C_1 and C_2 also prominent [22].

4

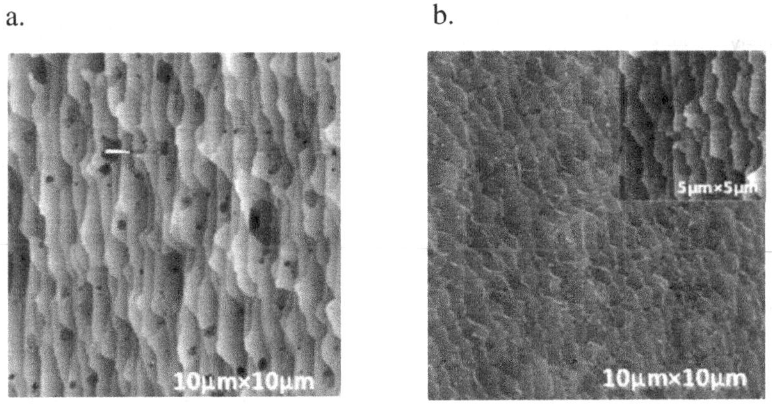

Fig. 1: AFM of graphene films grown at 1400°C for 30 min. a) sublimation grown graphene. b) C_{60} CMBE grown graphene at a flux of 8.7×10^{-8} Torr.

Prior to growth for both sources a tantalum film was deposited on the back of all samples for thermal management. After growth the samples were evaluated with Raman spectroscopy measurements. The thickness was estimated from X-ray photoelectron spectroscopy (XPS) measurements of the C1s peak intensity for graphene and SiC using the attenuation model of Seyller et al. [23]. Surfaces were studied with atomic force microscopy (AFM). Electrical measurements were made at room temperature on macroscopic indium contacted van der Pauw squares.

RESULTS AND DISCUSSION

Growth on Si-Face SiC

Raman measurements of CMBE grown films showed both the G and 2D bands in films grown at 1200°C with both sources [15] indicating the presence of graphene. It is usually assumed that temperatures above 1250°C are required for graphene growth by UHV sublimation on Si-face SiC [24] so this demonstrates lower temperature growth with CMBE. The disorder induced D band was weak or nonexistent in CMBE samples grown at 1400°C. Fitting of the 2D bands to multiple Lorentzians indicated multiple components for multilayer C_{60} material but only a single Lorentizian for multilayer GF material. This suggests that C_{60} material has Bernal stacking but GF material has random or turbostratic stacking. Figure 1 shows AFM images of graphene films grown by conventional sublimation and C_{60} CMBE at 1400°C for 30 min. The sublimation grown film is typical of graphene grown in UHV. The pit density is high and steps are irregular. The C_{60} CMBE sample is smoother with significantly lower pit density. The wrinkles common in multilayer films are clearly visible. GF CMBE films grown at 1200°C showed AFM similar to the 1400°C C_{60} CMBE shown in the figure. The AFM in ref. 10 also suggested a reduction in pit density.

Table I: Electrical Properties of sublimation grown and CMBE grown graphene films with typical growth conditions. T_{growth} and t_{growth} are the substrate temperature and growth time respectively.

Source	n/p (cm^{-2})	μ (cm^2/Vs)	Type	T_{Growth} (°C)	t_{Growth} (min)
Sublimation	4.5×10^{12}	256	n	1400	30
C_{60}	9×10^{12}	235	n	1400	30
GF	2×10^{12}	265	p	1200	60

Like Si-face sublimation grown material all C_{60} grown films showed n-type conduction. However, the GF material was consistently p-type, similar many to exfoliated and free standing graphene films. Typical sheet concentration and mobility values are given in Table I. In general the mobilities are low for the growth chamber used in these experiments.

As seen in fig. 2, the charge carrier sheet density for both sources depends linearly on film thickness, suggesting the doping coming from the source rather than the background pressure in the chamber. Experiments are underway to identify the source of the doping in both types of films.

Raman measurements have already

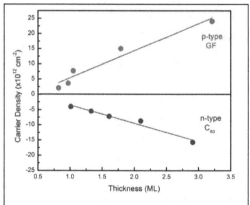

Fig. 2: Charge carrier sheet density vs. film thickness for n- and p-type CMBE samples.

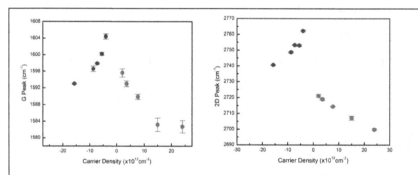

Fig. 3: Shift in Raman energy with doping for G and 2D bands.

6

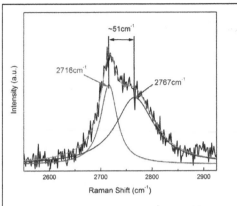

Fig: 4: Raman 2D for graphene stack with n-type C_{60} grown film on top of a p-type GF grown film.

been reported for both types of sources [15]. The SiC spectrum was subtracted in all spectra reported here. The D band was typically weak or not observable. There was a noticeable shift in peak position for the Raman bands with doping. Both of the Raman G and 2D bands show a phonon softening with increasing carrier density. This can be seen in fig. 3 where G and 2D peak positions are plotted as a function of sheet carrier density. A similar dependence of G band peak position versus charge density for gated graphene has been reported by Yan et al. [25].

While the doping effects reported here are accompanied with corresponding changes in thickness (fig. 2) they do suggest that with further refinement *in-situ* control of doping concentration and even type may be possible.

In addition to single carrier type growth, stacks of p doped and n doped graphene layers were made by first growing a film with the GF source then removing the sample, turning it over and then growing on top of the GF film with the C_{60} source. Figure 4 shows the Raman 2D for such a p-n stack. A high energy shoulder is clearly visible and fits show two peaks separated by 51 cm^{-1}. The individual components in Fig. 4 are similar to others reported in the literature so we therefore speculate that Fig. 4 represents two isolated layers of graphene with different doping. Fitting to the XPS C1s band (fig. 5) also indicates two separate films suggested by two graphene peaks. The splitting observed in both the Raman and XPS experiments is most likely due to different strain induced by two carbon sources. The GF growth rate is much faster than the C60 source and so the film might not be coupled to the SiC substrate in the same manner although interface layers are observed in the low energy regions of the XPS spectra for both types of material. Schmidt et al. [26] did report that the individual layers of folded exfoliated graphene

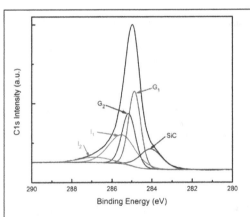

Fig. 5: XPS C1s band for p-n graphene stack showing splitting of the graphene component into two bands, G_1 and G_2.

monolayers were decoupled from each other and that even acted as separate conduction channels for electrical transport. The process of removing the sample to flip it may have caused the decoupling or it may be due to the different strain layers in the films with turbstratic graphene under Bernal stacked graphene.

Growth on Foreign Substrates

The ability to grow graphene directly on substrates other than SiC without transfer from other sources would be very beneficial though increasing the application space for graphene while reducing processing time and cost. There have been several reports of direct deposition of graphene on substrates other than SiC without the use of metal catalyst films. Hackley et al. [9] grew graphene like carbon on Si(111) substrates in an MBE chamber using e-beam evaporated graphite source. The confirmed the presence of graphene with XPS and Raman measurements. However, they did not show the region around the Raman 2D band and their D band was more intense than their G band. In addition to growth on C-face SiC Hwang et al. [12] recently demonstrated growth on sapphire substrates using a propane source but without metal catalysts. They too used Raman measurements but they had a strong well defined 2D peak and a low intensity ratio, I_D/I_G, indicating a low defect density and a quality similar to their material grown on C-face SiC. Synchrotron X-ray measurements indicated that the stacking order for multilayer films was predominantly rhombohedral rather than Bernal.

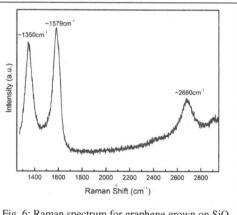

Fig. 6: Raman spectrum for graphene grown on SiO$_2$ by C$_{60}$ CMBE.

We report here the preliminary results of our studies of C$_{60}$ CMBE growth on alternate substrates. The most successful experiments involved growth on oxidized Si wafers, directly onto the SiO$_2$. Here the C$_{60}$ was deposited at 850°C for 30 min. then annealed at 1100°C for 30 min to form the graphene without evaporating the SiO$_2$. Figure 6 shows the Raman spectrum. A weak 2D band is present as well as the D and G bands. I_D/I_G is less than but close to one indicating significant defects and disorder. Very preliminary results suggest growth of graphene-like carbon films on GaN and sapphire substrates. The Raman results are similar to fig. 6 but with weaker 2D bands. Both G and D bands are strong.

CONCLUSIONS

Growth of graphene on SiC and other substrates by direct deposition of carbon without metallic catalysts has been demonstrated now by several laboratories. This is a viable process that provided several benefits over conventional SiC decomposition such as reduced pit density and lower growth temperatures. The technique eliminates the need for transfer of graphene from metal films to usable substrates as with CVD graphene. We have demonstrated CMBE growth of graphene in UHV from two different solid carbon sources, thermally evaporated C_{60} and a heated graphite filament. The carrier concentration, mobility and carrier type of the graphene films have been reported. GF grown films are consistently p-type while C_{60} grown films are n-type. Fitting of the Raman 2D bands suggests that the n-type C_{60} films have Bernal stacking while the p-type GF films are turbostratic. Growth of n-type on p-type stacks using both carbon sources was demonstrated. Raman and XPS measurements suggest that these stacks consist of two decoupled layers of graphene. We also report the growth of graphene-like carbon films on SiO/Si, sapphire and GaN substrates using C_{60} CMBE.

ACKNOWLEDGEMENTS

This work was supported by AFOSR (Dr. Harold Weinstock). The authors wish to thank Mr. Gerald Landis and Mr. John Hoelscher for technical assistance.

REFERENCES

1. K. S. Novoselov, A. K. Geim, S. V. Morozov, D. Jiang, Y. Zhang, S.V. Dubonos, I. V. Grigorieva, and A.A. Firov, *Science* **306**, 666 (2004).
2. A. E. Karu and M. Beer, *J. Appl. Phys.* **37**, 2179 (1966).
3. K. S. Kim, Y. Zhao, H. Jang, S. Y. Lee, J. M. Kim, K. S. Kim, J.-H. Ahn, P. Kim, J.-Y. Choi, and B. H. Kim, *Nature* **457**, 706 (2009).
4. X. Li, Y. Zhu, W. Cai, J. An, S. Kim, J. Nah, D. Yang, R. Piner, A. Velamakanni, I. Jung, E. Tutuc, S. K. Banerjee, L. Colombo and R. S. Ruoff, *Science* **324**, 1312 (2009).
5. S. Stankovich, D. A. Dikin, R. D. Piner, K. A. Kohlhaas, A. Kleinhammes, Y. Jia, Y. Wu, S. T. Nguyen and R. S. Ruoff, *Carbon*, **45**, 1558 (2007).
6. A. J. van Bommel, J. E. Crombeen, and A. van Tooren, *Surf. Sci.* **48**, 463 (1975).
7. C. Berger, Z. Song, X. Li, X. Wu, N. Brown, C. Naud, D. Mayou, T. Li, J. Hass, A. N. Marchenkov, E. H. Conrad, P. N. First, and W. A. de Heer, *Science* **312**, 1191 (2006).
8. K. V. Emtsev, A. Bostwick, K. Horn, J. Jobst, G. L. Kellog, L. Ley, J. L. McChesney, T. Ohta, S. A. Reshanov, J. Röhrl, E. Rotenberg, A. K. Schmid, D. Waldmann, H. B. Weber, and Th. Seyller, *Naturer Mater.* **8**, 203 (2009).
9. J. Hackley, D. Ali, J. DiPasquale, J. D. Demaree, and C. J. K. Richardson, *Appl. Phys. Lett.* **95**, 133114 (2009).
10. A. Al-Temimy, C. Riedl, and U. Starke, *Appl. Phys. Lett.* **95**, 231907 (2009).
11. E. Moreau, F. J. Ferrer, D. Vignaud, S. Godey, and X. Wallart, *Phys. Status Solidi A* **207**, 300 (2010).
12. J. Hwang, V. B. Shields, C. I. Thomas, S. Shivaraman, D. Hao, M. Kim, A. R. Woll, G. S. Tompa, and M. G. Spencer, *J. Cryst. Growth* **312**, 3219 (2010).
13. D. Usachov, V. K. Adamchuk, D. Haberer, A. Grüneis, H. Sachdev, A. B. Preobrajenski, C. Laubschat, and D. V. Vyalikh, *Phys. Rev.* **82**, 075415 (2010).
14. J. Park, W. C. Mitchel, L. Grazulis, H. E. Smith, K. G. Eyink, J. J. Boeckl, D. H. Tomich, S. D. Pacley, and J. E. Hoelscher, *Adv. Mater.* **22**, 4140 (2010).
14. W. C. Mitchel, J. H. Park, H. E. Smith, L. Grazulis, and K. Eyink, *Mater. Res. Soc. Symp. Proc.* **1246**, B10-02 (2010).
15. D. Chen, R. K. Workman, and D. Sarid, *J. Vac. Sci. Technol. B* **14**, 979 (1996).
16. E. Kolodney, B. Tsipinyuk and A. Budrevich, *J. Chem. Phys.* **100**, 8542 (1994).

17. A. V. Hamza, M. Balooch, and M. Moalem, *Surf. Sci.* **317**, L1129 (1994).

18. W. V. Lampert, C. J. Eiting, S. A. Smith, K. Mahalingham, L. Grazulis, and T. W. Haas, *J. Cryst. Growth* **234**, 369 (2002).

19. J. Li, P. Batoni, and R. Tsu, *Thin Solid Films* **518**, 1658 (2010).

20. R. J. Malik, R.N. Nottenberg, E. F. Schubert, J. F. Walker, and R. W. Ryan, *Appl. Phys. Lett.* **56**, 2651 (1988).

21. M. Joseph, N. Sivakumar, and P. Manoravi, *Carbon* **40**, 2031 (2002).

22. Th. Seyller, K. V. Emtsev, K. Gao, F. Speck, L. Ley, A. Tadlich, L. Broekman, J. D. Riley, R. C. C. Lackey, O. Rader, A. Varykhalov, and A. M. Shikhin, *Surf. Sci.* **600**, 3906 (2006).

23. J. Hass, W. A. de Heer and E. H. Conrad, *J. Phys.: Condens. Mattter* **20**, 1 (2008).

24. J. Yan, Y. Zhang, P. Kim, and A. Pinczuk, *Phys. Rev. Lett.* **98**, 166802 (2007).

25. H. Schmidt, T. Lüdtke, P. Barthold, E. McCann, V. I. Fal'ko, and R. J. Haug, *Appl. Phys. Lett.* **93**, 172108 (2008).

Mater. Res. Soc. Symp. Proc. Vol. 1284 © 2011 Materials Research Society
DOI: 10.1557/opl.2011.639

CVD Growth of Graphene on Three Types of Epitaxial Metal Films on Sapphire Substrate

Katsuya Nozawa, Nozomu Matsukawa, Kenji Toyoda and Shigeo Yoshii
Advanced Technology Research Laboratories, Panasonic Corporation, 3-4 Hikaridai, Seika,
Kyoto 619-0237, Japan

ABSTRACT

Graphene growth by chemical vapor deposition (CVD) was studied on three types of epitaxial metal films with different crystal structures on sapphire. Nickel (face-centered-cubic: fcc), Ru (hexagonal-closed-pack: hcp), and Co (fcc at temperature for graphene growth and hcp at R.T.) were deposited on c-face sapphire substrates and annealed in a furnace for solid phase epitaxial growth. Graphene layers were grown by CVD with methane gas on the epitaxial metal film.

The graphene layer uniformity was consistent with the structural simplicity of the metal film. The Ru sample had a single domain in the metal film and the highest graphene uniformity. The Co sample had a very complex crystal structure in the metal film and the poorest uniformity in graphene. The Ni sample had two types of stacking domains in the metal film and the graphene layer was uniform on each domain, but inhomogeneity was observed at domain boundaries.

INTRODUCTION

Graphene has been attracting a lot of attention [1] because of its remarkable properties, including high mobility [2] and long spin coherent length [3]. To utilize such properties in practical devices, a technique must be developed to grow a single domain and layer number controlled graphene films for each device on large substrates. Graphene growth on a metal surface is one candidate. Some metals work as catalysts for graphene growth [4]. High quality graphene films are synthesized by this method, but it is quite difficult to make a single metal substrate large enough for practical use.

Recently, experiments on chemical vapor deposition (CVD) growth of graphene on metal catalyst films have been reported [5, 6]. Some are sputtered metal films on amorphous layers such as SiO_2 [5], and others are freestanding metal foils [6]. It is easy to obtain large substrates for this method; even substrates as large as 30-inch are possible [6]. However, the metal films are poly-crystals. Crystal orientation of metal domain determines graphene orientation on it as shown in figure 1 (a). Grain boundary formation is unavoidable between graphene films grown on different orientation metal domains. In addition, grain boundary causes graphene layer number inhomogeneity [7].

Epitaxial growth of a metal catalyst film on a single crystal semiconductor substrate may solve this problem [8]. Some semiconductor substrates have similar lattice constant with catalyst metals and can have metal films epitaxially grown on them [9, 10]. Large semiconductor substrates are much easier to obtain than metal substrates of the same size. By using this method, an orientation controlled single domain metal surface may possibly be obtained for high quality graphene growth.

However, just using single crystal semiconductor substrates does not guarantee that a single metal crystal domain will form on them. Epitaxial growth of fcc metal such as Ni on a semiconductor tends to form two staking domains as shown in figure 1(b) [9]. On the other hand, Co has a phase transition point from hcp to fcc at 449 °C. Usually, graphene synthesis requires higher temperatures. Graphene is synthesized by using a metal lattice as a template, so the stacking domain formation and the phase transition may influence graphene synthesis. Nevertheless, no studies on them have been reported to the best of our knowledge.

In this paper, we tried epitaxial growths of three types of metals, Ni (fcc), Ru (hcp) and Co (fcc at temperature for graphene growth and hcp at R.T.) on sapphire substrate and CVD growth of graphene on them. We found that crystal structure of epitaxial metals plays a crucial role in graphene uniformity on them.

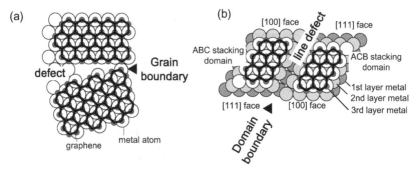

Figure 1. Grain boundary in poly-crystal (a) and domain boundary between two stacking domains in fcc (b).

EXPERIMENTAL DETAILS

Metal catalyst films were deposited on *c*-face sapphire substrates by RF plasma sputter in vacuum at room temperature. The thicknesses were around 150nm. The samples were annealed under hydrogen atmosphere at 1000°C for five minutes for solid phase epitaxy (SPE). After SPE, graphene layers were grown by the CVD method with the mixture of methane and hydrogen gas at about 1000°C. The metal films were characterized by a Nomarski microscope, X-ray diffraction (XRD), transmission electron microscope (TEM), and electron backscatter diffraction (EBSD). Graphene on the metal was characterized by TEM and a Raman microscope.

DISCUSSION

Crystal structure in metal films

Figure 2 shows XRD theta-2theta scans of metal films as deposition and after CVD. In all the cases, peaks from metal crystal plane parallel to sapphire surface are enhanced by the thermal process.

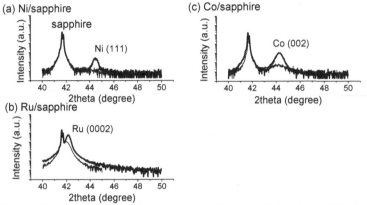

Figure 2. Theta - 2 theta scans of samples as deposition (solid line) and after CVD (bold line).

Figure 3 shows pole figures of reflection from crystal planes inclined to substrates. The sharp peaks mean that in-plane orientations of the metal crystals are well-defined by sapphire lattice. Both Ni and Ru show six similar peaks, but they have different meanings. The six peaks from Ni are from {200} planes. One fcc crystal domain has only three, so there are two domains in the Ni films. They are two stacking domains, like in figure 1(b).

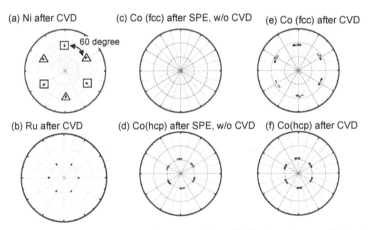

Figure 3. Pole figures of reflection from crystal planes inclined to substrates. Reflections from same domain are enclosed by same symbol in (a). No peaks are observed in (c).

On the other hand, the six peaks from Ru film come from {10$\bar{1}$3} planes. One hcp crystal has all six, which is evidence of single domain formation in the Ru film. The Co film has much more complex features than Ni or Ru. The sample after SPE without CVD growth shows only peaks from hcp structure, while samples cooled after CVD growth shows both peaks from {200} in fcc and {10$\bar{1}$3} in hcp. Carbon atoms diffused into Co film during CVD partially blocked phase transition from fcc to hcp. In addition, the number of peaks in the pole figure is not six, but twelve. There are two hcp and four fcc domains in the crystal after CVD.

TEM

Coherent growths of metal lattice on sapphire lattice and graphene growths on them are confirmed by TEM in figure 4. No amorphous layers are observed between the sapphire substrate and metal film, or between graphene and metal film. Some dislocations are noticed, but no grain boundaries are found in the field of view of Ni and Ru samples, while several grain boundaries are found in the field of view for Co sample.

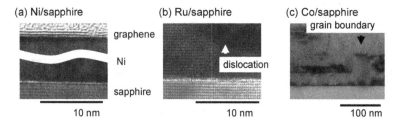

Figure 4. TEM pictures. Ni/sapphire (a), Ru/sapphire (b) and Co/sapphire (c)

Raman spectra

Figure 5 shows typical Raman spectra from graphene films on metals. All of them show features of few-layer

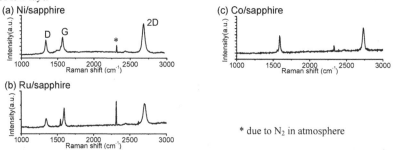

* due to N_2 in atmosphere

Figure 5. Raman spectra from graphene layers on metals.

All spectra were taken from samples with metal catalyst films. We found that underlying metal film affects Raman spectra. Relative intensities of G and 2D peaks to D peak are increased by removing metal film. Only the Ni sample shows a small peak around 1500 cm^{-1}. This peak also disappears when metal film is removed.

Raman mapping and EBSD analysis

Figure 6 shows Nomarski microscope images, orientation mapping by EBSD, and Raman intensity mapping. We found good coincidence between Nomarski microscope images and EBSD orientation mappings. Orientation mappings from two directions are shown only for the Co sample, because the Ni and Ru samples have simple crystal structures.

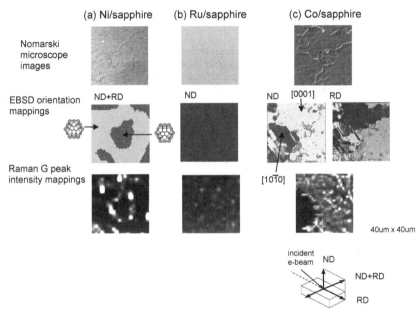

Figure 6. Nomarski microscope images, EBSD orientation mappings and Raman intensity mappings.

Two color regions in EBSD mapping of Ni sample correspond to two stacking domains in figure 1 (a). Intensity enhancement of G peak and decrease of 2D peak (not shown) are observed at the stacking domain boundary, while they are uniform on each domain. That is, the stacking domain boundary causes layer number inhomogeneity, like grain boundary in poly-crystal [7]. It is quite important to control stacking domains to form uniform graphene layers on epitaxial fcc metal film.

The Ru sample had a much smoother surface than other metals and the best uniformity in Raman spectra. This uniformity was due to its single domain nature. There were no grain or stacking domain boundaries, which cause inhomogeneity in graphene layers.

The crystal structure of the Co sample was very complex. Raman intensity was also highly inhomogeneous. EBSD analysis showed that the Co film consisted of a lot of small fcc and hcp domains facing in several directions. Some of the inhomogeneity in graphene layer number can be attributed to the difference in crystal face in ND direction, but there was also inhomogeneity that did not coincide with grain boundaries.

From these results, we can conclude that the structure of the metal film is crucial in graphene layer growth. Graphene layer uniformity is consistent with the simplicity of the metal structure.

CONCLUSIONS

This study shows for the first time the importance of crystal domain control in epitaxial metal film for graphene CVD growth. Not only grain boundary in poly-crystal but also stacking domain boundary in fcc crystal must be suppressed for uniform graphene growth. This study also demonstrated that using hcp metal, which is free from stacking domain problem, is one of the solutions for this problem.

REFERENCES

1. A.K.Geim and K.S.Novoselov, *Nature mat.*, **6**, 183 (2007).
2. K. I. Bolotin, K. J. Sikes, Z. Jiang, M. Klima, G. Fudenberg, J. Hone, P. Kim and H. L. Stormer, *Solid State Commun.*, **146**, 351 (2008).
3. N. Tombros, C. Jozsa, M. Popinciuc, H. T. Jonkman, B. J. van Wees, *Nature*, **448**, 571 (2007).
4. J. Wintterlin and M.-L. Bocquetb, *Surf. Sci.*, **603**, 1841 (2009).
5 . Y. Qingkai, J. Lian, S. Siriponglert, H. Li, Y. P. Chen, and S. -S. Pei, *Appl. Phys. Lett.*, **93**, 113103 (2008).
6. S. Bae, H. Kim,Y. Lee, X. Xu, J.-S. Park,Y. Zheng, J. Balakrishnan, T. Lei, H. R. Kim, Y. I. Song, Y.-J. Kim, K. S. Kim, B. Özyilmaz, J.-H. Ahn, B. H. Hong and S. Iijima, *Nat. Nanotechnol.*, **5**, 574 (2010).
7. Y. Zhang, L. Gomez, F. N. Ishikawa, A. Madaria, K. Ryu, C. Wang, A. Badmaev, and C. Zhou, *J. Phys. Chem. Lett.*, **1**, 3101 (2010).
8. H. Ago, I. Tanaka, M. Tsuji and K. Ikeda, *Small*, **6**, 1226 (2010).
9. H.Bialas and K.Heneka, *Vacuum*, **45**, 79 (1994).
10. S.Yamada, Y. Nishibe, H. Kitajima, S. Ohtsubo, A. Morimoto, T. Shimizu, K. Ishida and Y. Masaki, *Jpn. J. Appl. Phys.*, **41**, L206 (2002).

Low-Dimensional Graphitization and Structural Transformations

Mater. Res. Soc. Symp. Proc. Vol. 1284 © 2011 Materials Research Society
DOI: 10.1557/opl.2011.218

Low temperature CVD growth of graphene nano-flakes directly on high K dielectrics

Mark H. Rümmeli[1,2], Alicja Bachmatiuk[1], Arezoo Dianat[2], Andrew Scott[1], Felix Börrnert[1], Imad Ibrahim[1,2], Shasha Zhang[1], Ewa Borowiak-Palen[3], Gianaurelio Cuniberti[2,4] and Bernd Büchner[1]
[1]Leibniz-Institut für Festkörper- und Werkstoffforschung Dresden e. V., PF 27 01 16, 01171 Dresden,Germany
[2]Technische Universität Dresden, 01062 Dresden, Germany
[3]Zachodniopomorski Uniwersytet Technologiczny, Pulaskiego 10, 70322 Szczecin, Poland
[4]Division of IT Convergence Engineering and National Center for Nanomaterials Technology,POSTECH, Pohang 790-784, Republic of Korea

ABSTRACT

The potential of MgO and ZrO_2 as catalytically active substrates for graphene formation via thermal CVD is explored. Experimental observations show the growth of single and multi-layer graphene nano-flakes over MgO and ZrO_2 at low temperatures. The graphene nano-flakes are found to anchor at step sites. *Ab initio* calculations indicate step sites are crucial to adsorb and crack acetylene.

INTRODUCTION

Interest in graphene since its isolation in 2004 [1] has rapidly escalated and now with Geim and Novosalov being awarded the 2010 Nobel prize for physics, it looks set for even greater attention. It has been described as nature's thinnest elastic material and its exceptional mechanical and electronic properties make it an extremely exciting material. Within the realm of electronics, it is its one atoms thickness, planar geometry, high current-carrying capacity and thermal conductivity and potential to open a gap when existing as a narrow ribbon that hold particular promise. These features make it ideally suited for further miniaturizing electronics to form ultra-small devices and components for future semiconductor technology.

In order for graphene to realize its potential in electronics various obstacles need to be overcome. One of the more important aspects is its actual synthesis. Various routes exist to synthesize graphene however most are not best suited for integration in to current silicon technology. The primary routes are through graphite exfoliation, epitaxial graphene, graphene oxide and chemical vapour deposition. Most of these routes require the graphene be transferred onto a dielectric or, as in the case of SiC, require high temperatures.

In order to use them in field-effect transistors at room temperature one needs to modify graphene's semi-metallic nature so as to open a band gap. When existing as narrow strips (nanoribbons) quantum confinement effects lead to band gap formation [2]. Most band gap engineering routes use multiple lithographic steps to fabricate a graphene device. This leads contamination and disorder to the flake. Dry lithography-free techniques can help [3], but technical difficulties still remain. Another approach is chemical modification, for example, graphene oxide in which hydroxyl and other chemical groups attach to graphene. Although the technique is able to lift the degeneracy of the π band at the Fermi level of graphene, it is difficult to control its electronic properties and avoid defect formation [4]. Elias et al. [5] provided experimental evidence that the electronic structure of graphene can be modified by hydrogenation. Hydrogenation leads to the sp^2 bond type transforming to sp^3 type which removes the π bands and so opens a gap. This new material is known as graphane and was first predicted theoretically [6]. Another technical issue is the preparation of a gate when fabricating a graphene device. SiO₂ has inherent limits since its relatively low dielectric constant, K, leads to leakage currents quite easily. Top gate formation using high K dielectrics such as Al₂O₃ have been fabricated; however, this usually damages the graphene surface, see for example ref. [7]. The fabrication of graphene devices with atomically uniform gate dielectrics that can provide uniform electric field across the active region remains a challenge.

In this manuscript we discuss an unconventional route to form nano-sized graphene directly on high K dielectrics, namely MgO and ZrO₂. This is accomplished using standard thermal CVD techniques and can be achieved at temperatures below 500°C.

EXPERIMENTAL

The CVD reactions were conducted in a purpose built horizontal tube furnace with a sliding heating element which can be moved over the sample in an alumina boat. In the first set of experiments the oven was loaded with MgO nanopowder and evacuated to < 1 mbar and heated to a synthesis temperature of 775 °C. The synthesis was performed in a 100 mbar cyclohexane atmosphere and terminated after 17 s to 300 s by flushing with argon and sliding away the heating element so as to provide a rapid stop point. Further experiments were performed with MgO and ZrO₂ nano-crystalline powder using acetylene as the feedstock. After loading the samples the reactor was evacuated, then filled with 1015 mbar argon flow, and heated to temperatures ranging between 325 °C and 650 °C. The argon flow was held for 10 min and then switched to a mixture of acetylene and argon. The reaction was halted as discussed above. Some MgO samples were treated in dilute HCl to remove the MgO leaving only graphitic material. The microscropy investigations were performed on a FEI Titan[3] 80–300 aberration corrected electron microscope using an 80 kV acceleration voltage. The *ab initio* total energy calculations have been carried out by means of density-functional theory (DFT). They were performed within the PBE generalized gradient approximation for the exchange-correlation functional [8] and the projector augmented-wave method [9], using the Vienna *ab initio* simulation package (VASP) [10]. The activation energies were calculated using the nudged elastic band model developed by Henkelman's group [11]. The simulation cell included a slab of four substrate layers with a lateral size of a 4x4 surface unit cell which corresponds to 16 atoms on the surface layer. The step surface was prepared by removing eight atoms from the surface layer. In the case of an existing graphene on step edge surface, the lateral size of simulation cell was duplicated. In all simulations the vacuum gap between the slabs was larger than 15 Å. The wave functions have been expanded into plane waves up to a kinetic energy cut-off of 400 eV. The integration in the first Brillouin zone was performed using Monkhorst-Pack grids [12] including 25 k-points in the irreducible wedge. In all calculations, the two topmost layers of MgO have been optimized and from third layer up fixed in the bulk position until all force components were less than 0.01 eV/ Å. The convergence of energy differences with respect to the used cut-off energies and k-point grids has been tested in all cases within a tolerance of 10 meV/atom. The lattice constant of MgO is 2.10 Å and the spatial frequency in the reciprocal lattice of graphene is 2.13 Å

RESULTS & DISCUSSION

After the reaction the oxide powders, which are white initially, are visibly seen to have darkened. How dark they appear correlates with the reaction time such that those reacted for long periods are black while those reacted for time periods below 5 min become progressively paler as the reaction time is reduced.

Raman spectroscopy is an ideal tool to indicate the presence of sp^2 carbon through the G and D modes. All the samples made in this study were subjected to Raman investigations and all showed the presence of G and D modes typically with a ratio close to unity (data not shown) [13,14]. This ratio suggests a high degree of defects. However, many of these defects can be attributed to edge states [15]. Most importantly, the Raman data confirms the presence of graphitic carbon for all samples investigated.

Detailed TEM studies confirmed the presence of graphitic carbon. Samples formed through reaction times above 5 min. all showed graphitic layers on their surface (*e.g.* figure 1, left panel).

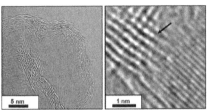

Figure 1. *Left panel*: TEM micrograph of graphitic layers encapsulating an MgO nano-crystal. *Right panel*: interface of graphitic layers with the MgO lattice fringes. Arrow indicates a graphene layer linking to a lattice fringe. (CVD parameters: precursor = cyclohaxane, temperature = 775°C, time = 300 sec.)

Remarkably all these samples (reaction time greater than 5 min.) all had the same number of layers ranging from 1 to 9 layers. This suggests the reaction stops once the oxide particles are fully encapsulated by graphitic layers. Interestingly all the graphitic layers can be traced to the crystal-graphene layer interface and were always found to anchor into step edges on the MgO (100) surface as for example highlighted in the right panel of figure 1. [14]. Previous work of ours suggests oxide supports play a role in the growth of multiwall carbon nanotubes [16, 17]. In these studies the graphitic layers were also found to anchor in the oxide support [17].

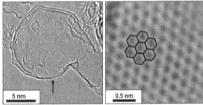

Figure 2 *Left panel*: TEM micrograph of graphitic shell after removal of MgO core with HCl treatment. Arrow points to single layer graphene region. *Right panel*: TEM micrograph taken from the single layer graphene region highlighted in the left panel.

Figure 2 shows a collapsed graphene jacket after the inner MgO nano-crystal was removed by dissolution in HCl. In one part of the collapsed structure a single layer of the membrane can be observed as indicated by the arrow in the right panel. Higher magnification of this single layer region confirms graphene (right panel).

As pointed out previously, we are able to tune the number of graphitic layers with reaction time. At very low reaction times (> 60 s) the number of layers can be driven to one. Often the surface of the MgO crystals contain graphene nano-flakes on their surface. At times a graphene layer can be seen to more or less cover an entire MgO nano-crystal.

In addition to magnesia, we also investigated the graphitization potential of zirconia. ZrO_2 is another high K dielectric and has also recently been extensively investigated as a catalyst for carbon nanotube growth [18]. In their studies they used in situ XPS studies to show zirconium oxide in a monoclinic baddeleyite or a slightly oxygen deficient form could form sp^2 carbon. Hence, one would anticipate graphitic growth under our reaction conditions using nano-crystallites. Post synthesis TEM lattice fringe studies confirmed the crystallites are in the baddeleyite form. In addition, the nano-crystals are covered with graphitic layers. Again the graphitic layers are observed to anchor to lattice fringes and in this case most appear to anchor on a face with (020) as its surface normal.

The common observation of graphitic layers terminating at step sites suggests they are important. Similar to observations in SiC [19] and previous studies of ours [16,17] indicate step sites initiate the growth of graphitic layers. Moreover, because growth apparently stops once a particle is fully encapsulated it is arguable these step sites are also involved in graphene nano-flake growth. To help elucidate these processes preliminary theoretical studies were conducted. Based on our experimental evidence we chose a step site on the (100) surface of MgO for our model system for graphene catalysis via oxides. Initially the adsorption and dissociation of acetylene on the MgO (100) surface and a step site was investigated. The left panel of figure 3 shows the 3 different adsorption sites investigated for a C_2H_2 molecule on the (100) surface (A, B & C) and on a step site (D, E & F). The adsorption energy is defines as the difference between the total energy of the system with the adsorbate (C_2H_2 molecule) and that without the adsorbate (referred to as the free energy of C_2H_2). The adsorption energies are shown in the middle panel of figure 5. The data shows C_2H_2 binds very weakly to the (100) surface. The distance between the surface and the molecules center of mass lies between 2.5 Å and 3 Å. These distances are in the physisorption adsorption regime. In contrast C_2H_2 binds more strongly at the step site. To investigate the dissociative adsorption energy of C_2H_2 to C_2H and H we use:

$$E_{DISS} = E_{C_2H+H^\cdot} - E_S - E_{C_2H_2}$$

Where $E_{C_2H+H^\cdot}$ is the total energy of C_2H and H on the substrate, E_S is the total energy of the clean substrate and $E_{C_2H_2}$ represents the free energy of acetylene in the gas phase. The calculated energies show the dissociative adsorption of C_2H_2 on the MgO (100) surface is an endothermic reaction whilst the reaction is exothermic on the step-edge surface. The data is shown in the right panel of figure 5.

The catalytic importance of step sites is well known. In the more specific case of catalytic sp^2 carbon formation with crystalline nickel [20, 21], sp^2 carbon growth can be divided into 4 sub-processes, namely, the adsorption of carbon precursor molecules on the catalyst, dissociation of H from the feedstock, surface diffusion and the addition of carbon atoms to the network. The above discussed DFT calculations confirm the adsorption and dissociation of the precursor on

MgO (100) is possible from step sites but not on the (100) surface. Thus step sites are critical to the process.

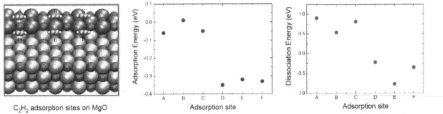

C₂H₂ adsorption sites on MgO   Adsorption site   Adsorption site

Figure 3 *Left panel*: The initial adsorption sites of C_2H_2 on (100) and step-edge of MgO surface. Calculated adsorption energies for the different sites (*middle panel*)and dissociative energies of C_2H_2 to C_2H and H (*right panel*).

We also investigated the diffusion barrier of carbon on MgO (100) surfaces. The total energy of the initial and final configurations were obtained using geometry optimization. The diffusion barrier for C was found to be ca. 0.38 eV, while that for an H atom on an oxygen atom is ca. 0.2 eV. This shows carbon can remain on the surface while H simply diffuses away. The studies also showed carbon adsorption on a MgO (110) surface with an existing graphene flake *viz.* carbon addition to a carbon honeycomb network [22].

SUMMARY

Our experimental studies show single and multi-layer graphene can be formed over ZrO_2 and MgO via CVD at temperatures below 500°C when using acetylene as the feedstock. Raman spectroscopic studies and TEM investigations confirm the formation sp^2 carbon. The number of graphene layers formed on the oxide nano-crystallites vary between 1 and 9. The number of layers can be tuned via the CVD reaction time. The TEM investigations revealed the graphitic layers are ubiquitously anchored at step edges.
DFT calculations were conducted on MgO (100) surfaces to better investigate the growth process. They clearly show step sites are essential for the adsorption and decomposition of C_2H_2. Diffusion studies suggest C remains on the surface while H simply diffuses away. In summary, our experimental and theoretical data indicate high K dielectrics have potential for the catalytic formation of graphene via thermal CVD at temperatures below 500°C.

ACKNOWLEDGMENTS

Computational resources were provided by the ZIH of the TU Dresden. A.B. acknowledges the A.-v.-Humboldt Stiftung and the BMBF, F.B. the DFG (RU 1540/8-1), S.Z. the IMPRS "Dynamical Processes in Atoms, Molecules and Solids", G.C. the South Korean Ministry of Education, Science, and Technology Program, Project WCU ITCE No. R31-2008-000-10100-0., and M.H.R. the EU (ECEMP) and the Freistaat Sachsen.

REFERENCES

1. K. S. Novoselov, A. K. Geim, S. V. Morozov, D. Jiang, Y. Zhang, S. V. Dubunos, I. V. Grigoriava and A. A. Firsov, Science 306, 666 (2004).
2. L. Jiao, L. Zhang, X. Wang, G. Diankov and H. Dai, Nature 458, 877 (2009).
3. M. Stanley, H. Wang, C. Puls, J. Forster, T. N. Jackson, K. McCarthy, B. Clouser and Y. Liu, Applied Physics Letters 90, 143518 (2007).
4. Y. H. Lu and Y.P. Feng, Journal of Physical Chemistry C 113, 20841 (2009).
5. D. C. Elias, R. R. Nair, T. M. G. Mohiuddin, S. V. Morozov, P. Blake, M. P. Halsall, A. C. Ferrari, D. W. Boukhvalov, M. I. Katsnelson, A. K. Geim and K. S. Novoselov, Science 323, 610 (2009).
6. J. O. Sofo, A. S. Chaudhari and G. D. Barber, Physical Review B 75, 153401 (2007).
7. M. C. Lemme, T. J. Echtermeyer, M. Baus and H. Kurz, IEEE Electron Device Letters 28, 282 (2007).
8. J. P. Perdew, K. Burke and M. Ernzerhof, Physical Review Letters 77, 3865 (1996).
9. P. E. Blöchl, Physical Review B 50, 17953 (1994).
10. G. Kresse and J. Furthmüller, Physical Review B 54, 11169 (1996).
11. G. Henkelman, A. Arnaldsson and H. Jonsson, Computational Materials Science 36, 354 (2006).
12. H. J. Monkhorst and J. D. Pack, Physical Review B 13, 5188 (1976).
13. M. H. Rümmeli, C. Kramberger, A. Grüneis, P. Ayala, T. Gemming, B. Büchner and T. Pichler, Chemistry of Materials 19, 4105 (2007).
14. M. H. Rümmeli, A. Bachmatiuk, A. Scott, F. Börrnert, J. H. Warner, V. Hoffmann, J. H. Lin,G. Cuniberti and B. Büchner, ACS Nano 4, 4206 (2010).
15. A.C. Ferrari, Solid State Communications 143, 47 (2007).
16. M. H. Rümmeli, F. Schäffel, C. Kramberger, T. Gemming, A. Bachmatiuk, R. J. Kalenczuk, B. Rellinghaus, B. Büchner and T. Pichler, Journal of the American Chemical Society 129, 15772 (2007).
17. M. H. Rümmeli, F. Schäffel, A. Bachmatiuk, D. Adebimpe, G. Trotter, F. Börrnert, A. Scott,E. Coric, M. Sparing, B. Rellinghaus, P. G. McCormick, G. Cuniberti, M. Knupfer, L. Schultz and B. Büchner, ACS Nano 4, 1146 (2010).
18. S. A. Steiner III, T. F. Baumann, B. C. Bayer, R. Blume, M. A. Worsley, W. J. MoberlyChan, E. L. Shaw, R. Schlögl, A. J. Hart, S. Hofmann and B. L. Wardle, Journal of the American Chemical Society 131, 12144 (2009).
19. K. V. Emtsev, A. Bostwick, K. Horn, J. Jobst, G. L. Kellogg, L. Ley, J. L. McChesney, T. Ohta, S. A. Reshanov, J. Röhrl, E. Rotenberg, A. K. Schmid, D. Waldmann, H. B. Weber and T. Seyller, Nature Materials 8, 203 (2009).
20. S. Hofmann, G. Csányi, A. C. Ferrari, M. C. Payne and J. Robertson, Physical Review Letters 95, 036101-1 (2005).
21. S. Helveg, Clopez-Cartes, J. Sehested, P. L. Hansen, B. S. Clausen, J. R. Rostrup-Nielsen,F. Abild-Pedersen, and J. K. Nørskov, Nature (London) 427, 426 (2004).
22. A. Scott, A. Dianat, F. Börrnert, A. Bachmatiuk, S. Zhang, J. H. Warner, E. Borowiak-Palen, M. Knupfer, B. Büchner, G. Cuniberti and M. H. Rümmeli, Applied Physics Letters. submitted.

Mater. Res. Soc. Symp. Proc. Vol. 1284 © 2011 Materials Research Society
DOI: 10.1557/opl.2011.640

On the carbo-thermal reduction of silica for carbon nano-fibre formation via CVD

Alicja Bachmatiuk[1], Felix Börrnert[1], Imad Ibrahim[1], Bernd Büchner[1], and Mark H. Rümmeli[1,2]

[1]IFW Dresden, P.O. Box 270116, 01069 Dresden, Germany
[2]Technische Universität Dresden, 01062 Dresden, Germany

ABSTRACT

The formation of carbon nanostructures using silica nanoparticles from quartz substrates as a catalyst in an aerosol assisted chemical vapor deposition process was examined. The silica particles are reduced to silicon carbide via a carbothermal reduction process. The recyclability of the explored quartz substrates is also presented. The addition of triethyl borate improves the efficiency of the carbothermal reduction process and carbon nanotubes formation. Moreover, the addition of hydrogen during the chemical vapor deposition leads to the helical carbon nanostructures formation.

INTRODUCTION

It is well known that the graphitization process for carbon nanomaterials formation mostly use metal catalysts [1-6]. Which remain as impurities after the synthesis reaction. These metal impurities are for many applications unwanted and so strategies to eliminate or avoid them are being developed. The most popular method to remove metal impurities is through the use of post-synthesis chemical treatments, e.g. by acids [7, 8]. However, these treatments, on the whole, are not able to fully eliminate the metals and additionally such treatment can be destructive to the carbon structures themselves. Thus, there is interest to replace metal catalyst systems through non-metal synthesis routes. The implementation of the ceramic catalysts (SiO_2, MgO, ZrO_2, Al_2O_3, SiC) for the synthesis of carbon materials is a rapidly developing field [9-20]. In particular, the utilization of SiO_2 for carbon nanomaterials synthesis is very attractive for further application in silicon based technology. In this article we present metal free synthesis of sp^2 carbon nanomaterials from silicon carbide particles formed via carbothermal reduction of silica. The addition of triethyl borate as an acceleration agent for the carbon nanomaterials synthesis and hydrogen as a modifying carbon structure medium were also investigated. Moreover, the recycling potential for the substrates is also demonstrated.

EXPERIMENT

The chemical vapor deposition experiments were accomplished in a horizontal oven using an aerosol system to introduce the carbon source [9, 10]. The carbon nanostructures were formed at the centre of the furnace over pristine quartz substrates as well as over previously used substrates. The CVD experiments were performed using a temperature of 900 °C and the reaction time was 0.5 hour. The carbon source: ethanol or an ethanol/triethyl borate mixture was transported to the reactor via flowing argon (7200ml/min) through the spray injection system which forms a fine mist. In addition, nominal additions of H_2 were investigated using the same flow rates. The recyclability processes consisted of two steps: initial burning of the previously formed material in the air at 800 °C and then washing the substrate surfaces with distilled water and drying. The produced nanomaterials were characterized using electron microscopy (scanning electron microscopy – SEM and transmission electron microscopy – TEM) and spectroscopy techniques (optical absorption and Raman). The SEM measurements were performed on FEI Nova-Nanosem. The TEM investigations were conducted using aberration corrected microscopes: FEI Titan[3] operated at 80 kV and Jeol 2010F operated at 200 kV. Optical absorption measurements (OAS) were conducted on a Bruker 113 Fourier transform spectrometer and Raman measurements using a Thermo Scientific DXR SmartRaman spectrometer ($\lambda = 532$ nm).

DISCUSSION

Figure 1 presents SEM overview images of the produced carbon nanostructures using pure ethanol feedstock and (a) pristine quartz substrate, (b) using previously used quartz substrate after synthesis with ethanol (recycled substrate), (c) using a ethanol/triethyl borate feedstock mixture and pristine quartz substrate, (d) using pure ethanol and 35 at. % hydrogen addition over a previously used quartz substrate after synthesis with ethanol/triethyl borate mixture. When the pristine ethanol and fresh quartz substrate were used the surface of the substrate after the reaction was for the most part covered with an amorphous carbon layer [9]. Some small graphitic humps could be also observed.

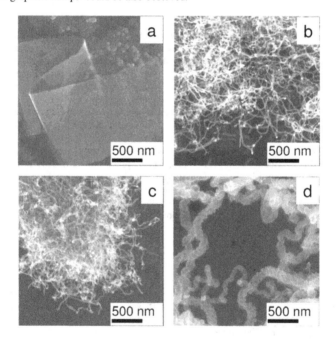

Figure 1. Scanning electron microscopy images of the produced carbon nanostructures **a** – using pure ethanol feedstock and a pristine quartz substrate, **b** – using pure ethanol and previously used quartz substrate after the synthesis with ethanol, **c** – using ethanol/triethyl borate mixture feedstock and pristine quartz substrate, **d** – using pure ethanol and 35 at. % of hydrogen addition over previously used quartz substrate after synthesis with ethanol/triethyl borate mixture.

In the case of using pure ethanol as the feedstock and a previously used quartz substrate, after the synthesis with the formation of nanotubes in dense mats can easily be observed (Figure 1b). When an ethanol/triethyl borate feedstock mixture with a pristine quartz substrate is used, the formation of dense mats of nanotubes is observed without the need to use a recycled substrate. Figure 1c shows an example of a dense mat of CNT. This suggests the influence of the boron addition on the pristine quartz substrate is rather dramatic [9, 21]. The boron oxide formed from the thermal decomposition of triethyl borate reduces the melting temperature of quartz causing the surface changes [21]. The surface after the experiment with boron was noticeably rougher and more nanosized particles were formed for potential nanotubes growth. This suggests boron oxide accelerates a process crucial to CNT formation. In the experiments where H_2 (35 at. %) was added into the spray injector gas line, significantly different structures were obtained. The fibers were mostly much thicker than those formed in the previous experiments and appear to have a coiled structure as can be seen in figure 1d.

The higher resolving power of transmission electron microscopy as compared to SEM makes it a useful technique to investigate the morphology of the formed carbon nanostructures in more detail. Figure 2 presents representative TEM images of the obtained nanotubes. For fibers formed in the experiments without hydrogen, the structures had diameters ranging from 10 nm to 35 nm and consisted of stacked cup graphitic layers (e.g. figure 2 a). Many of the fibers have holes at their poles so that pockets form in the core of the fiber (e.g. figure 2b). The interlayer spacing of the graphitic caps is around 0.35 nm.

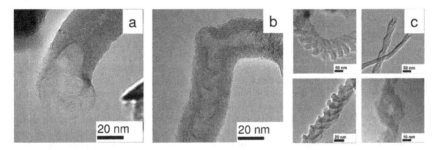

Figure 2. Transmission electron microscopy images providing the closer investigation of the produced carbon nanostructures **a** – using pure ethanol and previously used quartz substrate after synthesis with ethanol, **b** – using ethanol/triethyl borate mixture feedstock and pristine quartz substrate, **c** – using pure ethanol and 35 at. % of hydrogen addition over previously used quartz substrate after synthesis with ethanol/triethyl borate mixture.

The TEM studies confirm the SEM investigations which suggested that the addition of 35 at. % of hydrogen to the leads to the formation of helical nano-fibers. Various types of helical or

27

twisted structures can be observed. Some examples are presented in figure 2c. Many of the helical structures appear to form through inter-twining behavior of two or more fibers. The yield of helical structures as estimated through TEM observations to be around 95 %. The diameter distribution of the helical structures ranged from 10 to 150 nm.

Spectroscopic measurements were conducted in order to better comprehend the processes occurring in the reactions. In order to confirm the presence of sp^2 carbon formation the Raman measurements were conducted. The appearance of the typical signals around 1350 cm^{-1} (Defect mode) and 1600 cm^{-1} (Graphitic mode) during the measurements were observed (figure 3a). The quality of the samples was determined by the ratio of the G to D modes (G/D ratio). The highest G/D ratio was measured for the sample produced using pure ethanol and a pristine substrate – *solid line* (0.77). The next best G/D ratio occur using ethanol/triethyl borate mixture feedstock and pristine quartz – *dotted line* (0.74). The samples produced from pure ethanol and previously used quartz substrate after synthesis with ethanol – *dashed line* and using pure ethanol and 35 at. % of hydrogen addition over previously used quartz substrate after synthesis with ethanol/triethyl borate mixture, have G/D ratios around 0.72.

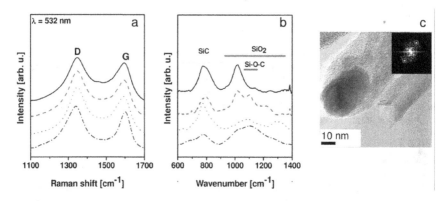

Figure 3. a – Optical absorption spectra from the produced samples. **b** – Raman spectra showing sp^2 carbon formation. *Solid line* – using pure ethanol feedstock and a pristine quartz substrate, *dashed line* – using pure ethanol and previously used quartz substrate after synthesis with ethanol, *dotted line* – using ethanol/triethyl borate mixture feedstock and pristine quartz substrate, *dashed dotted line* – using pure ethanol and 35 at. % of hydrogen addition over previously used quartz substrate after synthesis with ethanol/triethyl borate mixture. **c** – TEM image of the SiC particle at the base of a carbon nanotube. Inset: FFT of (100) hexagonal SiC particle.

Figure 3 b presents typical OAS measurements where three responses refer to the presence of SiC, SiO_2, and Si-O-C. The silicon dioxide signal arises from the starting catalyst material. The silicon carbide signal comes from the reduction of SiO_2 during the CVD process via a carbothermal reaction [9, 10]. The signal from the presence of silicon oxycarbide is related to the partial conversion of silica to silicon carbide. With the addition of hydrogen the silicon carbide peak is weaker what can be related to its less efficient formation and possible amorphization [22]. The SiO_2 and Si-O-C signals are broadened with a use of H_2 which can be related to an increase in amorphous Si species.

The presence of SiC fits with TEM data in which particles at the root of a fiber (e.g. figure 3c) or in the core of a fiber consistently showed the presence of SiC. This suggests growth occurs from SiC. Thus, one would anticipate SiC signatures in the OAS data. The carbo-thermal reduction of silica to SiC is described by Weimer et al. [23], Vix-Guterl et al. [24] and Bachmatiuk et al. [9]. In short, reactions conducted in pure ethanol reduce silica to SiC slowly and hence recycled substrates are more efficient as they were previously partially reduced. The presence of B species in the reaction apparently accelerates the carbo-thermal reduction process. The manner with which this happens has yet to be fully clarified. The addition of hydrogen to the reaction leads to the formation of helical structures. It is proposed that hydrogen radicals etch the sides of the stacked cup fibers where under normal circumstances the layers link to each other. Hydrogen radicals etch these links, leaving open layers so that when neighboring fibers touch, the layers cross link driving an inter-twinning process which ultimately leads to helical fibers [25].

CONCLUSIONS

We have shown that the formation of stacked-cup carbon nanofibres is possible through a chemical vapor deposition process using SiO_2 as non-metallic catalyst. The SiO_2 is reduced to silicon carbide through a carbothermal reduction process. TEM studies show SiC nano-particles at the roots of the formed stacked cup nano-fibers indicating their growth occur from SiC rather than SiO_2. The addition of hydrogen to the synthesis leads to the formation of helical structures. This is attributed to the etching ability of hydrogen radicals which open connected stacked cup layers in the fibers. Thus, when these open ended layers come into contact with adjacent layers from a neighboring tube they link. This begins a cross-linking intertwining self-assembling process that produces helical nanostructures.

ACKNOWLEDGMENTS

A.B. acknowledges the A.-v.-Humboldt Stiftung and the BMBF, F.B. the DFG (RU 1540/8-1), II thanks the DAAD (A/07/80841) and MHR the EU (ECEMP) and the Freistaat Sachsen.

REFERENCES

1. T. Baker, Chem Ind (Lond) 18, 698 (1982).
2. H. Yoshida, S. Takeda, T. Uchiyama, H. Kohno, Y. Homma, Nano Letters 8, 2082 (2008).
3. S. Costa, E. Borowiak-Palen, A. Bachmatiuk, M. H. Rümmeli, T. Gemming, R. J. Kalenczuk, Phys Status Solidi B 244, 4315 (2007).
4. A. Bachmatiuk, E. Borowiak-Palen, M. H. Rümmeli, T. Gemming, R. J. Kalenczuk, Phys Status Solidi B 244, 3925, (2007).
5. O. C. Carneiro, M. S. Kim, J. B. Yim, N. M. Rodriguez, R. T. K Baker, Journal of Physical Chemistry B 107, 4237 (2004).

6. M. H. Rümmeli, F. Schäffel, A. Bachmatiuk, D. Adebimpe, G. Trotter, F. Börrnert, ACS Nano 4, 1146 (2010).
7. A. Bachmatiuk, E. Borowiak-Palen, M. H. Rümmeli, C. Kramberger, H-. W. Hübers, T. Gemming, Nanotechnology 18, 275610 (2007).
8. A. Bachmatiuk, E. Borowiak-Palen, R. J. Kalenczuk, Nanotechnology, 19, 365605 (2008).
9. A. Bachmatiuk, F. Börrnert, M. Grobosch, F. Schäffel, U. Wolff, A. Scott, M. Zaka, J. H. Warner, R. Klingeler, M. Knupfer, B. Büchner, M. H. Rümmeli, ACS Nano, 3, 4098 (2009).
10. A. Bachmatiuk, F. Börrnert, F. Schäffel, M. Zaka, G. Simha Martynkowa, D. Placha, R. Schonfelder, P. M. F. J. Costa, N. Ioannides, J. H. Warner, R. Klingeler, B. Büchner, M. H. Rümmeli, Carbon, 48, 3175 (2010).
11. M. H. Rümmeli, A. Bachmatiuk, A. Scott, F. Börrnert, J. H. Warner, V. Hoffmann, J. H. Lin,G. Cuniberti and B. Büchner, ACS Nano 4, 4206 (2010).
12. S. A. Steiner III, T. F. Baumann, B. C. Bayer, R. Blume, M. A. Worsley, W. J. MoberlyChan, E. L. Shaw, R. Schlögl, A. J. Hart, S. Hofmann and B. L. Wardle, Journal of the American Chemical Society 131, 12144 (2009).
13. D. Takagi, H. Hibino, S. Suzuki, Y. Kobayashi, Y. Homma Nano Letters 7, 2272 (2007).
14. D. Takagi, Y. Kobayashi, Y. Homma, Y. Carbon nanotube growth from diamond. Journal of the American Chemical Society 131, 6922 (2009).
15. Liu B, RenW, Gao L, Li S, Pei S, Liu C, et al. Metal-catalyst-free growth of single-walled carbon nanotubes. Journal of the American Chemical Society 131, 2082 (2009).
16. M. H. Rümmeli, E. Borowiak-Palen, T. Gemming, T. Pichler, M. Knupfer, M. Kalbac, Nano Letters 5, 1209 (2005).
17. M. H. Rümmeli, C. Kramberger, A. Grüneis, P. Ayala, T. Gemming, B. Büchner B, Chemistry of Materials, 19, 4105 (2007).
18. S. Huang, Q. Cai, J. Chen, Y. Qian, L. Zhang, Journal of the American Chemical Society 131, 2094 (2009).
19. A. Hirsch, Angewandte Chemie International Edition, 48, 5403 (2009).
20. H. Liu, D. Takagi, S. Chiashi, Y. Homma, Carbon 48, 114 (2010).
21. J. Hlavac, Glass science and technology, the technology of glass and ceramics, 12 Elsevier: Czechoslovakia; (1983).
22. Y. Sun, T. Miyasato, Japanese Journal of Applied Physics 37, 5485 (1998).
23. A. W. Weimer, K. J. Nilsen, G. A. Couchran, R. P. Roach, AIChE Journal, 39, 493 (1993).
24. C. Vix-Guterl, I. Alix, P. Gibot, P. Ehrburger, Applied Surface Science, 210, 329 (2003).
25. A. Bachmatiuk, F. Börrnert, V. Hoffmann, D. Lindackers, J-H. Lin, B. Büchner, M. H. Rümmeli, European Journal of Chemistry, submitted.

Mater. Res. Soc. Symp. Proc. Vol. 1284 © 2011 Materials Research Society
DOI: 10.1557/opl.2011.641

A Fully Atomistic Reactive Molecular Dynamics Study on the Formation of Graphane from Graphene Hydrogenated Membranes.

Pedro A. S. Autreto[1] , Marcelo Z. Flores[1], Sergio B. Legoas[2], Ricardo P. B. Santos[1,3] and Douglas S. Galvao[1]

[1] Instituto de Física "Gleb Wataghin, Universidade Estadual de Campinas, Campinas - SP, 13083-970, Brazil

[2] Departamento de Física, CCT, Universidade Federal de Roraima, 69304-000, Boa Vista - RR, Brazil.

[3] Departamento de Engenharia Agrícola, Universidade Estadual de Maringá, 82020-900, Maringá - PR, Brazil.

ABSTRACT

Recently, Elias et al. (Science **323**, 610 (2009).) reported the experimental realization of the formation of graphane from hydrogenation of graphene membranes under cold plasma exposure. In graphane, the carbon-carbon bonds are in sp^3 configuration, as opposed to the sp^2 hybridization of graphene, and the C–H bonds exhibit an alternating pattern (up and down with relation to the plane defined by the carbon atoms). In this work we have investigated, using reactive molecular dynamics simulations, the role of H frustration (breaking the H atoms up and down alternating pattern) in graphane-like structures. Our results show that a significant percentage of uncorrelated H frustrated domains are formed in the early stages of the hydrogenation process, leading to membrane shrinkage and extensive membrane corrugations. This might explain the significant broad distribution of values of lattice parameter experimentally observed. For comparison purposes we have also analyzed fluorinated graphane-like structures. Our results show that similarly to H, F atoms also create significant uncorrelated frustrated domains on graphene membranes.

INTRODUCTION

The discovery of new carbon-based materials has been frequent in recent decades. Examples of these materials are colossal nanotubes [1] and graphene [2]. Graphene is a two dimensional array of hexagonal units of sp^2 bonded C atoms with very unusual and interesting electronic and mechanical properties [2]. Because of its electronic properties, graphene is considered one of the most promising materials for future electronics [3]. However since graphene is a gapless material, its use becomes restrict in some electronic applications [4]. One possibility towards opening graphene gap is through chemical functionalization, using hydrogen or fluorine atoms [5-14].

Fully hydrogenated graphene, named graphane, was theoretically predicted by Sofo, Chaudhari, and Barber [5], and experimentally realized by Elias et al. [7]. In their experiments, graphene membranes were submitted to H$^+$ exposure from cold plasma. The H incorporation into the membranes results in altering the carbon hybridizations from sp^2 to sp^3. The experiments were also made with graphene membranes over SiO$_2$ substrates, producing a material with different properties [7].

Perfect idealized graphane consists of a single-layer structure with fully saturated (sp^3 hybridization) carbon atoms and with C-H bonds in an alternating pattern (up and down with relation to the plane defined by the carbon atoms). Its two most stable conformations are the so called chair-like (H atoms on both sides of the plane) and boat-like (H atoms alternating in pairs) [5]. These structures are shown in Fig. 1.

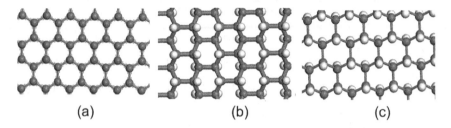

(a)   (b)   (c)

Figure 1: Structural models. Carbon and hydrogen atoms are indicated in grey and white colors, respectively. (a) graphene, (b) graphane boat-like, and; (c) graphane chair-like.

MODELING

In this work we have investigated, through molecular dynamics (MD) simulations, the structural and dynamical aspects of the hydrogenation and fluorination of graphene membranes, leading to the formation of graphane and fluorographene-like structures. We carried out extensive MD calculations using reactive force fields (ReaxFF [15-17]), as implemented in the Large-scale Atomic/Molecular Massively Parallel Simulator (LAMMPS) code [18]. We have considered different temperature regimes (300, 500, and 650 K) and monoatomic gas atmospheres (hydrogen and fluorine). The MD simulations were realized employing different percentages of hydrogen and fluorine (H/C and F/C), ranging from 0.1 to 2.0. The objective was to verify how different atmosphere conditions could affect the H and F incorporation processes. We used finite (no cyclic boundary conditions) structures with H-passivated borders. The membranes were embedded in a gas of randomly distributed H and F atoms (typical size of 30 x 30 unit cells). In order to speed up the simulations, the H-H and F-F recombinations were not permitted. The Berendsen thermostat, as implemented in LAMMPS code, was used and the typical time for a complete simulation run was 100 ps, with time-steps of 1fs.

ReaxFF is a reactive force field developed by van Duin, Goddard III and co-workers for use in MD simulations. It allows simulations of many types of chemical reactions. It is similar to standard non-reactive force fields, like MM3 [19], where the system energy is divided into partial energy contributions associated with, amongst others, valence angle bending and bond stretching, and non-bonded van der Waals and Coulomb interactions [15-17]. However, one main difference is that ReaxFF can handle bond formation and dissociation (making/breaking bonds) as a function of bond order values. ReaxFF was parameterized against DFT calculations, being the average deviations between the heats of formation predicted by the ReaxFF and the experiments equal to 2.8 and 2.9 kcal/mol, for non-conjugated and conjugated systems, respectively [15-17]. We have carried out geometry optimizations using gradient conjugated techniques (convergence condition with gradient values less than 10^{-3}).

RESULTS AND DISCUSSIONS

We started analyzing the incorporation rate of F and H as a function of their concentrations and temperatures. In Fig. 2 we show typical results for the number of formed C-H and F-C bonds, as a function of the simulation time. Due to its higher chemical reactivity, at the initial states the number of formed F-C bonds is almost one order of magnitude higher that C-H ones. After 100 ps, the incorporation rates become very similar: 0.88 atom/ps and 0.74 atom/ps for hydrogen and fluorine, respectively. This can be explained by the fact that as the time passes, the number of available sites for F becomes much smaller than the available ones to H. Increasing the temperature and percentage of hydrogen just resulted in faster H incorporation in the graphene membranes. The situation is completely different in the case of fluorine, where the high temperature and high percentage of fluorine create holes in the membranes and can partially destroy them.

In Fig. 2 we show typical results for the time evolution of the number of formed H-C and F-C bonds; for the case of H/C and F/C ratios close to 1.0, and temperature equals to 500 K. At the beginning, the graphene is a planar structure in a hydrogen atmosphere (Fig. 3a). H incorporation generates a corrugated structure as showed in the Fig. 3b. The results show that a significant percentage of uncorrelated H frustrated domains is formed in the early stages of the hydrogenation processes. These results also suggest that large domains of perfect graphane-like structures are unlikely to be formed. A frustration (Fig.3c) is a configuration where the necessary graphane sequence of alternating up and down H atoms is no longer possible (i.e., become frustrated) [19] (see Figs. 3 and 4).

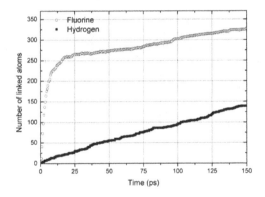

Figure 2: Hydrogen and fluorine incorporation on graphene layers as a function of time. Results from calculations carried out at 500 K and with H/C and F/C ratios approximately equal to 1.0.

Our results show that H frustration domains are very stable, as high temperatures are needed to reverse graphane-like structures to their original graphene configurations. Higher temperatures lead to an increase in the rate of incorporation and on uncorrelated domains formation, which can increase the number of frustrations. These aspects will be discussed in details in a forthcoming publication [20]. The results for F are quite similar, with the main differences that at high temperatures and F gas density, F creates holes in the membranes and can partially destroy it [20].

We have also investigated the role of the frustration in the geometry of the graphene membranes. Two different H frustrations were considered here, one with parallel H atoms (Frust-1) and the other with missing H atoms (Frust-0) (see Fig. 4).

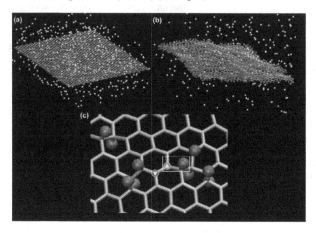

Figure 3: (a) Representative snapshots of the early hydrogenation stage from ReaxFF molecular dynamics simulations. Non-bonded H atoms are indicated in white and C-bonded ones in red. (b) Final stage of hydrogenation. In panel (c) is showed a representative snapshot of the final hydrogenation state. We observed the formation of diverse frustrated hydrogenated domains. In (c), the arrows indicate the way how an H atom would be bonded to a C atom, up (red arrow) or down (blue arrow). The yellow rectangle highlights one possible frustration state.

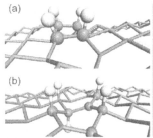

Figure 4. Examples of different possible frustration types. (a) Frust-1, H frustration with parallel first-neighbor H atoms; (b) Frust-0, H frustration with 'missing' first-neighbor H atoms. Atoms in the defect region are shown in a ball and stick rendering. For clarity the H atoms outside this region were made transparent.

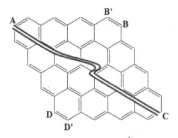

Figure 5: Schematic drawing of graphene fragment before hydrogenation. The letters are the reference points fôr the distances displayed in Table 1. The diagonal bi-line (joining A and C points) is just to indicate that the size of the membranes is bigger than showed here.

Table 1: Distances calculated with ReaxFF, in Angstroms, between the reference points for the system depicted in Fig. 5. G-chair refers to the chair-like graphane system. Frust-0 and Frust-1 refer to defects which were described in text and shown in Fig. 4. The numbers in parentheses indicate the number of frustrated domains in the structure. Frust-1-in-line and Frust-0-in-line refer to frustration (of type 1 and 0, respectively) created along a line through the graphene membrane. The graphane-in-graphene refers to a small region of graphene structure in graphene membrane.

System	d_{A-B}	$d_{B'-C}$	d_{C-D}	$d_{D'-A}$	d_{A-C}	d_{B-D}
Graphene	22.26	22.26	22.26	22.26	37.08	22.47
G-chair	22.95	22.94	22.95	2.94	38.40	23.01
Frust-1(13)	22.21	22.92	22.80	22.90	37.95	22.47
Frust-0(13)	22.38	22.98	22.77	22.82	38.12	22.37
Frust-1-in-line	22.82	20.93	22.98	22.92	37.84	22.81
Frust-0-in-line	23.00	21.22	23.41	22.98	35.12	22.34
Graphene-in-graphene	22.15	22.07	21.93	22.07	36.58	22.26

We have measured some representative distances (indicated in Fig. 5) in order to quantify how different hydrogenation patterns affect the geometry of the graphene membranes. This allows us to evaluate the level of shrinkage and corrugation of the hydrogenated graphene fragments. As we can see from Table 1 and Figure 3, the H frustration increases out-of-plane distortions, which induce in-plane geometrical shrinkage. This effect is amplified when Frust-1 is present. For particular configurations these distances can even be smaller than the corresponding graphene values [7]. Similar effects, but with more pronounced distortions, were observed for fluorinated structures [20].

SUMMARY AND CONCLUSIONS

We have investigated using reactive molecular dynamics methods the hydrogenation and fluorination of graphene membranes. Our simulations show that H and F frustrations are very likely to occur. A significant percentage of uncorrelated H and F frustrated domains are easily formed in the early stages of the hydrogenation/fluorination process leading to decreased lattice values and extensive membrane corrugation. Our geometry optimization calculations, involving finite graphane-like membrane, show that H and F frustration increases out-of-plane distortion, which induces in-plane dimensional shrinkage.

ACKNOWLEDGMENTS

This work was supported in part by the Brazilian agencies CNPq, CAPES and FAPESP. The authors wish to thank Professor A. van Duin for his very helpful assistance with ReaxFF code.

REFERENCES

[1] H. Peng et al., Phys. Rev. Lett. 101, 145501 (2008).
[2] K. S. Novoselov et al., Science 306, 666 (2004).
[3] S. H. Cheng et al., Phys. Rev. B 81, 205435 (2010).
[4] F. Withers, M. Duboist, and A.K. Savchenko, arxiv:1005.3474v1 (2010).
[5] J. Sofo, A. Chaudhari, and G. Barber, Phys. Rev. B 75, 153401 (2007).
[6] S. Ryu et al., Nano Lett. 8, 4597 (2008).
[7] D. Elias et al. Science 323, 610 (2009).
[8] J. O. Sofo, A. S.Chaudhari, and, G. D. Barber, Phys. Rev. B 75, 153401 (2007).
[9] D. Lueking et al., J. Am. Chem. Soc. 128, 7758 (2006).
[10] N. R. Ray, A. K. Srivastava, and, R. Grotzsche, arXiv:0802.3998v1 (2008).
[11] O. Leenaerts, H. Peelaers,A. D. Hernandez-Nieves, B. Partoens,and F. M. Peeters, , arxiv:1009.3847v1 (2010).
[12] S.-H. Cheng, K. Zou,F. Okino, H. R. Gutierrez, A. Gupta, N. Shen, P. C. Eklund, J. O. Sofo, and J. Zhu, Phys. Rev. B 81, 205435 (2010).
[13] R. R. Nair et al., Small, in press, DOI: 10.1002/smll.201001555.
[14] J. T. Robinson et al., Nano Lett., in press, DOI: 10.1021/nl101437p.
[15] A. C. T. van Duin, S. Dasgupta, F. Lorant, and W. A. Goddard III, J. Phys. Chem. A 105, 9396 (2001).
[16] A. C. T. van Duin and J. S. S. Damste, Org. Geochem. 34, 515 (2003).
[17] K. Chenoweth, A. C. T. van Duin, and W. A. Goddard III, J. Phys. Chem. A 112, 1040 (2008).
[18] http://lammps.sandia.gov/
[19] M. Z. S. Flores, P. A. S. Autreto, S. B. Legoas, and D. S. Galvao, Nanotechnology 20, 465704 (2009).
[20] R. B. P. Santos, P. A. S. Autreto, S. B. Legoas, and D. S. Galvão, to be published.

Poster Session: Mechanism, Growth, and Processes of Low-Dimensional Carbon Nanostructures

Mater. Res. Soc. Symp. Proc. Vol. 1284 © 2011 Materials Research Society
DOI: 10.1557/opl.2011.219

Metal-catalyzed graphitization in Ni-C alloys and amorphous-C/Ni bilayers

Katherine L. Saenger, Christian Lavoie, Roy Carruthers, Ageeth A. Bol, Timothy J. Mcardle, Jack O. Chu, James C. Tsang, and Alfred Grill
IBM Semiconductor Research and Development Center
Research Division, T. J. Watson Research Center, Yorktown Heights, NY 10598

ABSTRACT

Metal-catalyzed graphitization from vapor phase sources of carbon is now an established technique for producing few-layer graphene, a candidate material of interest for post-silicon electronics. Here we describe two alternative metal-catalyzed graphene formation processes utilizing solid phase sources of carbon. In the first, carbon is introduced as part of a cosputtered Ni-C alloy; in the second, carbon is introduced as one of the layers in an amorphous carbon (a-C)/Ni bilayer stack. We examine the quality and characteristics of the resulting graphene as a function of starting film thicknesses, Ni-C alloy composition or a-C deposition method (physical or chemical vapor deposition), and annealing conditions. We then discuss some of the competing processes playing a role in graphitic carbon formation and review recent evidence showing that the graphitic carbon in the a-C/Ni system initially forms by a metal-induced crystallization mechanism (analogous to what is seen with Al-induced crystallization of amorphous Si) rather than by the dissolution-upon-heating/precipitation-upon-cooling mechanism seen when graphene is grown by metal-catalyzed chemical vapor deposition methods.

INTRODUCTION

Few-layer graphene has attracted intense interest as a possible material for post-silicon electronic devices due to its high mobility, two-dimensional structure, and tunable band gap [1-3]. Methods for forming graphene such as mechanical exfoliation from graphite [3] and decomposition of single-crystal SiC [4] are not readily scalable to the wafer-scale dimensions that are expected to be required for semiconductor manufacturing. One potentially scalable method is metal-catalyzed chemical vapor deposition (CVD), in which graphene is formed on a metallic template layer exposed to a carbon-containing gas at elevated temperature (900-1000 °C). Several groups have shown that it is possible to grow few-layer graphene on Ni and transfer it to insulating substrate layers [2,5,6].

We have been investigating alternative metal-catalyzed graphene formation processes utilizing solid-phase sources of carbon. In a first approach, shown in Fig. 1A, the carbon is introduced as a component of a Ni-C alloy film; in a second, the carbon is introduced as a layer in an amorphous carbon (a-C)/Ni bilayer stack [7,8], as shown in Fig. 1B. It was hoped that these approaches might provide films of quality comparable to those achieved by CVD, but with better control over film thickness (since the carbon supply is fixed and finite.

In this work, we examine the quality and characteristics of the graphene produced with these two methods as a function of starting film thicknesses, Ni-C alloy composition or a-C deposition method, and annealing conditions. We then discuss some of the competing processes playing a role in graphitic carbon formation and review recent evidence [8] showing that the graphitic carbon in the a-C/Ni system initially forms by a metal-induced crystallization mechanism (analogous to what is seen with Al-induced crystallization of amorphous Si) rather than by the dissolution-upon-heating/precipitation-upon-cooling mechanism seen when graphene is grown by metal-catalyzed chemical vapor deposition methods.

EXPERIMENT

All substrates were thermally oxidized 100-oriented Si wafers (SiO$_2$ thickness ~300 nm). For the alloy experiments, Ni-C films with carbon concentrations of approximately 1, 3, and 10 at% percent were deposited by cosputtering from separate Ni and C targets. For the bilayer experiments, the a-C was deposited by physical vapor deposition (PVD) or plasma-enhanced chemical vapor deposition (PECVD). For samples with PVD a-C, substrates were *in situ* sputter precleaned and then sputter deposited with a layer of a-C (3, 10, or 30 nm) followed (without breaking vacuum) by a layer of Ni (30, 100, or 300 nm). For the bilayer experiments with PECVD a-C, samples were coated at 550 °C with one of three thicknesses (7, 12, or 32 nm) of a-C (from a C-containing precursor diluted in He) in a first tool and transferred to a second tool for a room temperature in-situ sputter preclean (estimated to remove about 2 nm of carbon) followed by sputter deposition of 100 nm of Ni.

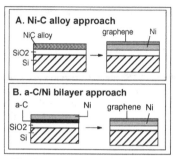

Fig. 1. Two approaches for metal-catalyzed formation of graphitic C in schematic cross section view.

Heat treatments typically consisted of rapid thermal anneals (RTAs) with 35 °C/s heating rates to 650 - 1000 °C in N$_2$ or Ar/H$_2$(5%) for times ranging from 10 s to 1 min. The default cooling rate was uncontrolled (>>10 °C/sec), but was sometimes slowed to 2, 5, or 10 °C/sec. Samples were characterized either as-grown or after Ni etching (in HCl) and transfer to a Si/SiO$_2$(300 nm) substrate by a "release and catch" or "in-place bonding" [9] method, where any protective PMMA layers remaining after in-place bonding were selectively removed by annealing in air or O$_2$ at 400-450 °C for 3 - 30 min. Samples were characterized by optical and scanning electron microscopy, profilometry (for the thicker films), 4-point probe sheet resistance (Rs), Raman spectroscopy (with 514.5 or 532 nm laser excitation and spatial resolution of 1-2 μm), and x-ray diffraction (XRD) around the 002 reflection of graphitic carbon (corresponding to a d-spacing of 0.34 nm) with θ–2θ scans in a Bragg-Brentano geometry with Cu K$_\alpha$ radiation (λ = 0.1542 nm).

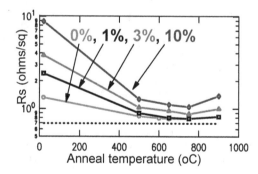

Fig. 2. Rs for 100-nm-thick Ni-C films with 0, 1, 3, and 10 at % C after 1 min anneals (lines); calculated Rs for a 100 nm film with the 6.8 μΩ-cm resistivity of bulk Ni (dotted).

RESULTS AND DISCUSSION

Ni-C alloy

Heat treatments of the as-deposited Ni-C alloy films were expected to produce both grain growth and a redistribution of carbon, both of

40

which affect film sheet resistance. Figure 2 shows the room temperature Rs values for 100-nm-thick Ni and Ni-C alloy films on Si/SiO$_2$ substrates after 1 min anneals in N$_2$ as a function of anneal temperature. The data reflect two trends: an increase in Rs with increasing C content (in part due to alloy scattering effects), an effect especially apparent in the as-deposited films, and a decrease in Rs with

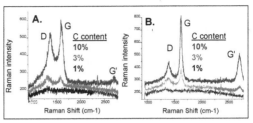

Fig. 3. Raman spectra of graphitic carbon produced by annealing Si/SiO$_2$/Ni-C(100 nm) samples for 1 min in N$_2$ at 750 °C (A) or 900 °C (B).

increasing annealing temperature (a Ni grain growth effect). The large decrease in Rs in the 10% C film (factor of 7) after the 500 °C anneal suggests that the heat treatments are also decreasing the within-grain carbon content, with the expelled carbon precipitating into a form that does not interfere with grain-to-grain conduction. The slight increase in Rs between 750 and 900 °C might reflect an increase in dissolved C, or to a mild agglomeration effects reducing film continuity.

The Raman data of Fig. 3 confirm that graphitic carbon is present on the top surface of the NiC films after annealing at 750 or 900 °C for 1 min in N$_2$. As expected, Raman

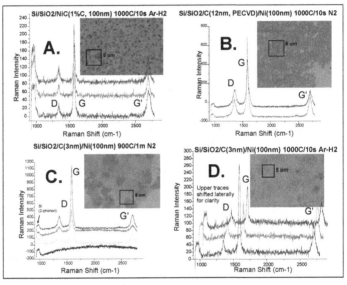

Fig. 4. A sampling of Raman results and optical images for graphitic carbon films produced from Ni-C alloys or a-C/Ni bilayers after in-place bonding transfer to Si/SiO$_2$ substrates. Films from: Ni-C(1% C) after 1000 °C/10s (10 °C/s cool to 700 °C) annealing in Ar-H$_2$ (A); a-C(7 nm PECVD)/Ni (100 nm) after 1000 °C/10 s (2 °C/s cool to 700 °C) annealing in N$_2$ (B); a-C(3nm, PVD)/Ni(100 nm) after 900 °C/1m annealing in N$_2$ (C); and a-C(3 nm, PVD)/Ni(100 nm) after 1000 °C/10 s annealing (10 °C/s cool to 700 °C) annealing in Ar-H$_2$ (D). Different traces in the same plot are for different positions; intensity offsets introduced for clarity.

intensity increases with the initial C content of the film. The relative intensities of the D, G, and G' peaks at ~1350, ~1580, and ~2700 cm^{-1} indicate that the carbon is highly disordered after the 750 °C anneal, but has the signature of few or multilayer graphene (FLG or MLG) after the 900 °C anneal. Figure 4A shows an optical image and Raman data (3 positions) for one of the better films produced by this technique (a Ni-C film with 1% C annealed at 1000 °C for 10 s in Ar/H$_2$ with a controlled cooling to 700 °C at 10 °C/s) after transfer to the underlying Si/SiO$_2$ growth substrate by "in-place" bonding using a protective PMMA layer subsequently removed by a 450 °C/3min air bake. The relatively high G/D ratios (~2 -10), moderate G'/G ratios (~1 to 4), and relatively weak intensity suggest a film of FLG. The speckled morphology in the optical image of Fig. 4A is quite typical for both the 1 and 3 % films, whereas the 10% films typically showed similar speckles at a much higher areal density.

<u>a-C/Ni Bilayer</u>
Graphitic carbon was produced from a-C/Ni bilayer samples having a wide range of a-C and Ni thicknesses. Figure 5 compares optical, SEM, and XRD data for a-C/Ni(100nm) samples with three initial thicknesses of PVD a-C (3, 10, and 30 nm) after 900 °C/1 min N$_2$ annealing. The presence of surface carbon suggested by the optical images is more clearly shown in SEM. The 3 nm sample shows distinct, somewhat overlapping thin patches; the 10 nm sample shows a thicker, nearly continuous surface layer; and the 30 nm sample looks thick and wrinkled. XRD shows detectable graphitic carbon in all three samples, with higher a-C thicknesses producing 002 graphite peaks with higher intensities and narrower widths. Similar optical images and XRD intensities were obtained for samples with PECVD a-C when matched for a-C thickness and annealing conditions.

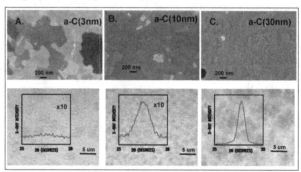

Fig. 5. Comparison of optical, SEM, and XRD data for bilayer Si/SiO$_2$/a-C/Ni(100 nm) samples after 900 °C/1 min N$_2$ annealing.

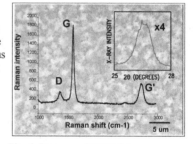

Fig. 6. Raman and XRD data superimposed on an optical micrograph of a transferred MLG film produced by annealing Si/SiO2/a-C(10 nm)/Ni(100 nm) in N$_2$ at 900 °C for 1 min.

Figure 6 superimposes Raman and XRD data on an optical image of a 10-12 nm thick (measured by profilometry) graphitic carbon film prepared by 900 °C/1m N$_2$ annealing of a a-C(10nm)/Ni(100nm) bilayer after transfer to a Si/SiO$_2$ substrate by the release-and-catch method. The 002 XRD peak remains strong and is

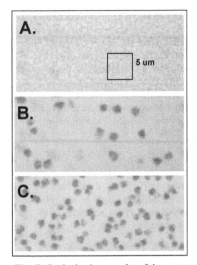

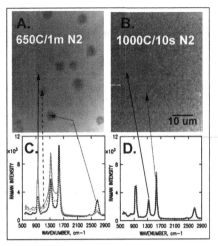

Fig. 7: Optical micrographs of three areas of same Si/SiO₂/a-C(10 nm)/Ni(100 nm) sample after 650 °C/1 min. N₂ annealing.

Fig. 8. Optical image-correlated Raman data for C films samples derived from Si/SiO₂/a-Si(3 nm)/Ni (100 nm) after transfer to Si/SiO₂ substrates. Left (A, C): a sample annealed at 650 °C for 1 min in N₂; right (B, D): a sample annealed at 1000 °C for 10 s in N₂.

similar in shape to what it was on the original substrate. The G/G' ratio of ~4 and relatively high (~10) G/D ratio suggests a reasonable quality MLG film.

Raman data and optical images for three of our better films are shown in Figs. 3B - 3D.

Mechanism Issues

A concern with the bilayer a-C/Ni process is the highly variable MLG morphology. Figure 7 shows that substantial variability can be present even within different regions of the same sample for the case of a bilayer Si/SiO2/a-C(10nm, PVD)/Ni(100nm) sample annealed at 650 °C for 1 min in N₂. In some regions of the sample, the precipitated carbon appears relatively fine-grained and homogeneous (Fig. 7A), whereas other regions show ~1 μm diam dark flakes with a distribution that is sparse (Fig. 7B) or dense (Fig. 7C). An inhomogeneous morphology appears to correlate with more variety in the Raman data, as can be seen from the data of Fig. 8 for two carbon films on Si/SiO₂ substrates transferred by in-place bonding after annealing Si/SiO₂/a-C(3nm, PVD)/Ni(100 nm) samples in N₂ at 650 °C for 1 min or 1000 °C for 10 s. The 650 °C film of Fig. 8A has an inhomogeneous morphology and very position-dependent Raman spectra (Fig. 8C). In contrast, the 1000 °C film of Fig. 8B has a more uniform morphology (at least on the scale of the ~1 μm regions sampled by Raman) and very similar Raman spectra (Fig. 8D) over the entire film area.

We speculate that some of this nonuniformity may be a consequence of the competing processes taking place during early stages of annealing, namely C-Ni intermixing and Ni grain growth. Previous *in situ* XRD studies [8] have established that the graphitic carbon in the a-C/Ni system initially forms *during heating* by a metal-induced crystallization mechanism (analogous to what is seen with Al-induced crystallization of amorphous Si [10]), but that the graphitic carbon formed can also dissolve back into the Ni upon heating to higher temperatures and precipitate out upon cooling (as seen with Ni-catalyzed CVD). Both of these mechanisms rely on the steps of (i) mass transport of C atoms into the Ni (from the initial a-C or the graphitic carbon formed during heating), (ii) diffusion of C within (and/or on) the Ni, and (iii) graphitic C precipitation and growth at nucleation sites. However, in contrast to CVD methods, the Ni grain growth in the bilayer a-C/Ni system occurs concurrently with carbon incorporation, rather than in a separate step [6] before the Ni is exposed to a gaseous carbon source. Since the C-content of Ni will affect Ni grain growth, slight differences in the local thermal environment at early times might be expected to have a large effect on the final Ni grain structure and C morphology.

SUMMARY

We have demonstrated metal-catalyzed graphene formation with solid phase sources of carbon introduced as either (i) part of a co-sputtered Ni-C alloy or (ii) one of the layers in an amorphous carbon (a-C)/Ni bilayer stack. Reasonable quality multilayer graphene was produced for a wide variety of conditions, but concerns remain about inhomogenities in sample morphology, especially with films in the thickness range of interest (<3 nm).

ACKNOWLEDGMENTS

This work was supported by DARPA under contract FA8650-08-C-7838 through the CERA program. We thank C.-Y. Sung for management support, the Microelectronics Research Laboratory staff for their contributions to sample preparation, and T. Graham, A. Pyzyna, K. Sivakumar, and D.A. Neumayer for help with processing and process development.

REFERENCES

1. A.K. Geim and K. S. Novoselov, Nature Materials **6**, 183 (2007).
2. K.S. Kim, Y. Zhao, H. Jang, S.Y. Lee, J.M. Kim, K.S. Kim, J.-H. Ahn, P. Kim, J.-Y. Choi, and B.H. Hong, Nature **457**, 706 (2009).
3. K.S. Novoselov, A.K. Geim, S.V. Morozov, D. Jiang, Y. Zhang, S.V. Dubonos, I.V. Grigorieva, and A.A. Firsov, Science **306**, 666 (2004).
4. C. Berger, Z. Song, T. Li, X. Li, A.Y. Ogbazghi, R. Feng, Z. Dai, A.N. Marchenkov, E.H. Conrad, P.N. First, and W.A. de Heer, J. Phys. Chem. B **108**, 19912 (2004).
5. Q. Yu, J. Lian, S. Siriponglert, H. Li, Y.P. Chen, and S.-S.Pei, Appl. Phys. Lett. **93**, 113103 (2008).
6. A. Reina, X. Jia, J. Ho, D. Nezich, H. Son, V. Bulovic, M.S. Dresselhaus, and J. Kong, Nano Lett. **9**, 30 (2009).
7. M. Zheng, K. Takei, B. Hsia, H. Fang, X. Zhang, N. Ferralis, H. Ko, Y.-L. Chueh, Y. Zhang, R. Maboudian, and A. Javey, Appl. Phys. Lett. **96**, 063110 (2010).
8. K.L. Saenger, J.C. Tsang, A. Bol, J.O. Chu, A. Grill, and C. Lavoie, Appl. Phys. Lett. **96**, 153105 (2010).
9. A.A. Bol, J.O. Chu, A. Grill, C.E. Murray, K.L. Saenger, U. S. Patent No. 7,811,906 (12 October 2010).
10. O. Nast and S.R. Wenham, J. Appl. Phys. **88**, 124 (2000); O. Nast and A.J. Hartmann, J. Appl. Phys. **88**, 716 (2000).

Mater. Res. Soc. Symp. Proc. Vol. 1284 © 2011 Materials Research Society
DOI: 10.1557/opl.2011.642

Hole or Electron Doped C_{60} Polymer Using Free Electron Laser Irradiation

Nobuyuki Iwata, Daiki Koide, Syouta Katou, Eri Ikeda, and Hiroshi Yamamoto
CST, Nihon Univ., 7-24-1 Narashinodai, Funabashi-shi, Chiba 274-8501, Japan

ABSTRACT

Polymerized C_{60} crystals were grown using the free electron laser (FEL) irradiation. In order to promote the polymerization degree, hole or electron was doped in the C_{60} crystals grown by the liquid-liquid interfacial precipitation (LLIP) method to eliminate the degradation by oxidation. The specimen grown with the I_2 dissolved butylalcohol (BTA, $CH_3(CH_2)_3OH$) and the C_{60} saturation o-xylene solution, subsequently pressed at 7GPa, showed only Ag(2)-derived mode at 1456 cm^{-1} after the FEL irradiation. The specimen belonged to so-called F phase, which is not obtained by the typical photo-induced polymerization process. It was noted that the FEL irradiation for polymerization of C_{60} was quite useful.

INTRODUCTION

The three dimensional (3D) C_{60} polymer can be obtained by high-presure and high-temperature (HPHT) method [1]. The sub-mm 3D crystalline C_{60} polymer is prepared by HPHT from single crystal of two dimensinal (2D) C_{60} polymers [2]. In the 3D-C_{60}, C_{60} moleculres are polymerized via [3+3] cycloadition in addition to the [2+2] one [3]. For example, otrhorhombic 2D-C_{60} polymaer is converted to orthorhombic 3D-C_{60} polymer under 15GPa at 500-700ºC, however, the 30% 2D-C_{60} remains and a graphite-like amorphous phase is obtained at temperatures higher than 700ºC. The aim of our study is to develop a novel photopolymerization process for C_{60} molecules and synthesize an amorphous 3D C_{60} polymer at the bulk scale for various application fields. The C_{60} polymer is expected to have features of stiffness, low-density as well as flexibility like organic material owing to the amorphous phase of C_{60} polymer, and the density lower than that of metal alloys [4]. We chose the FEL as a light source, which has unique features: a tunable wavelength in the infrared range, and a micropulses with a pulse width less than one ps [5]. H. Nakayama et $al.$ demonstrated that a hole doping in graphite induces the same effect as applying pressure [6]. Whereareas, the polymerization is carried out by doping an electron and a hole using a sccaning tunnneling microscope (STM) tip, meaning that both of the doping is effective to promote the polymerization, where the polymerization degrree depends on the molecules distance [7-9]. In our previous report, in the specimen with shorter molecular distance under 7GPa, the polymerization degree is also promoted, and we found the hole doping is effective for polymerization. However, the preparation of the specimen of mixed C_{60} and iodine was done in air, where a degradation of molecules is expected by oxidation [10,11]. Futhermore the polymerization area is limited to approximately 5µm in diameter when the FEL is irradiated to the pressed C_{60} powder. The reason is probably the limited directions for polymerization. The directionality is expected to be reduced inserting some molecules with double bonds between C_{60} molecules and then an amorphous C_{60} polymer is also anticipated [12]. In this study the hole or electron doped C_{60} crystals were grown by the LLIP method not to exposure the specimens to the air, and then the FEL was irradiated to the pressed supecimens in the saturated C_{60} solution [13].

EXPERIMENTAL PROCEDURES

The C_{60} crystals were grown by the LLIP method. A saturated C_{60} solution with o-xylene, and the I_2 dissolved for hole doping and calcium methoxide (CMO), $Ca(COH_3)_2$, saturated for electron doping BTA were prepared. The I_2 powder, 0.393 mg, was dissolved in the 20 ml BTA, and the CMO, 0.3 g was added to the 100 ml BTA to be supersaturation state. The solution was sonicated for 15-30 min. In the case of the CMO, saturation solution was used. The C_{60} crystals were grown by two different kinds of procedures by the LLIP. First, the I_2 dissolved or the CMO saturated BTA 10-20 ml was gently added on the C_{60} saturation solution set in a beaker, where the ratio of the BTA solution to the C_{60} saturation solution was 2 : 1. The two layered solution was maintained in a refrigerator at approximately 10°C for one week. Second, the two layered solution was sonicated for 15-30 min, and then it was maintained in a refrigerator at approximately 10°C for one day, or the solution was left at room temperature for one week until the solvent was completely evaporated. In both cases, if the specimens were wet, the solvent was removed. The remnant grown C_{60} crystals in a beaker was dried for two days in dark case not to be oxidized with light. The dried C_{60} crystals were pressed at 600MPa or 7GPa in air, and then the specimens were soaked in C_{60} saturation solution with the height of specimen surface and the solution surface same. The 4th harmonic 500 nm-FEL of 2000 nm fundamental wavelength with 7 mm in diameter was irradiated to the specimen surface with 0.1-1.0mJ/cm^2/pulse at 2Hz for one hour. The surface was analyzed by micro-Raman spectroscopy with low power density of 1.27 W/mm^2 to exclude the polymerization by the analysis.

RESULTS & DISCUSSION

Figures 1(a) and 1(b) show the Raman spectra around the Ag(2) pentagonal pinch mode and Ag(1) radial breathing mode, respectively. The pristine C_{60} mono-molecule shows the Ag(2) at 1469 cm^{-1} and the Ag(1) at 493 cm^{-1}. The measured peak position of the Ag(2) was always in the range from 1469 to 1469.5 cm^{-1}, indicating the resolution of 0.5 cm^{-1} in this equipment. All Raman spectra illustrated below are normalized with the maximum peak intensity in the described range in each figure. The specimen was grown with the I_2 dissolved BTA and the C_{60} saturation o-xylene solution in a refrigerator, and then pressed at 600MPa. The dotted lines are of before the FEL irradiation, and the solid lines are of after the FEL irradiation. The dotted line shows the Ag(2) peak at 1469 cm^{-1}, which was similar to that of pristine C_{60} as shown in (a). In addition to the Ag(1) at 493 cm^{-1}, the Ag(1)-derived mode at 473 cm^{-1} was observed, which was indicative of the polymerization in part in the specimen as shown in (b). The small peak of Ag(2)-derived mode at 1456 cm^{-1} in (a) as well as the Ag(1)-derived mode at 473 cm^{-1} described with solid lines indicated also the polymerization in part.

Figure 2 shows the Raman spectra of the specimen, which was grown with the BTA and the C_{60} saturation o-xylene solution in a refrigerator, and then pressed at 7GPa. In both cases, the peaks at Ag(2) mode at 1469 cm^{-1}, and Ag(2)-derived mode at 1458 cm^{-1} were observed before the FEL irradiation in (a). After the FEL irradiation, the peak intensity of Ag(2) mode was much reduced, and the derived mode at 1457 cm^{-1} was promoted noted by solid arrow in (a). Only the Ag(1)-derived modes at 473 cm^{-1} were obtained in both cases.

In Ref.[1,14], at room temperature and 7GPa, the orthorhombic phase is realized, where the C_{60} polymer via [2+2] cycloaddition chained. Therefore even though before the FEL irradiation, the polymerization peak by the Ag(2)-derived mode at 1458 cm^{-1} was observed.

Whereas, the polymerization degree was promoted after the FEL irradiation, demonstrated by the increasing of the Ag(2)-derived mode intensity, noted by the solid arrow.

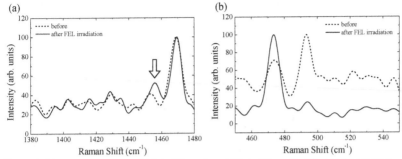

Figure 1. Raman spectra around (a) Ag(2) mode and (b) Ag(1) mode. The specimens were grown with the I_2 dissolved BTA and the C_{60} saturation o-xylene solution. The specimens were pressed at 600MPa. The appearance of Ag(2)-derived mode at 1456 cm^{-1} noted by arrow and Ag(1)-derived mode indicated the polymerization in part described by solid lines in (a) and (b).

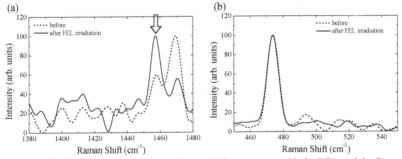

Figure 2. The Raman spectra of the specimen, which was grown with the BTA and the C_{60} saturation o-xylene solution in a refrigerator. The specimen was pressed at 7GPa. The polymerization degree was promoted by the FEL irradiation illustrated in (a). In both cases, the Ag(1)-derived mode was observed at 473 cm^{-1}.

The Raman spectra of the specimen grown with the I_2 dissolved BTA and the C_{60} saturation o-xylene solution in a refrigerator are shown in figure 3. The specimen was pressed at 7GPa. Before the FEL irradiation, the Ag(2)-derived mode at 1458 cm^{-1}, denoted by dotted arrow, was observed in addition to the Ag(2) mode at 1469 cm^{-1}. After the FEL irradiation, the Ag(2) mode at 1469 cm^{-1} completely shifted to the 1456 cm^{-1} as shown in (a). In (b) only peaks at 473 cm^{-1} were observed.

The same discussion, mentioned above, can be applied to the appearance of the Ag(2)-derived mode before the FEL irradiation because of the 7GPa pressure. It is worthy to mention that the FEL irradiation transferred the partial polymerization state into complete polymerization state, demonstrated by the disappearance of Ag(2) mode at 1469 cm^{-1} and appearance of Ag(2)-derived mode at 1456 cm^{-1} in (a).

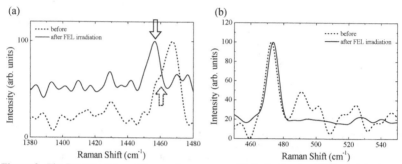

Figure 3. The Raman spectra of the specimen grown with the I_2 dissolved BTA and the C_{60} saturation o-xylene solution in a refrigerator. Since the specimen was pressed at 7GPa, polymerized orthorhombic phase is anticipated even though the FEL is not irradiated. The Ag(2)-derived mode was observed, meaning the partial polymerization, denoted by dotted arrow in (a). The peak was completely moved downward, described by solid arrow in (a). In the Ag(1) as shown in (b), both the specimens indicated the polymerization.

Figure 4 shows the Raman spectra of the specimen, which was grown by the sonicated mixture solution of the CMO saturated BTA and the C_{60} saturation o-xylene solution, and then left it in a refrigerator for one day. The grown specimen was pressed at 600MPa. The Ag(2)-derived mode at 1457 cm^{-1} denoted by solid arrow indicated the presence of partial polymerization phase as shown in (a). From the results of the Ag(2) in (a) and the Ag(1) in (b), as-grown C_{60} crystals were not polymerized.

When the specimens were grown using the I_2 dissolved BTA and C_{60} saturation solution with sonication distribution, the remarkable results were not obtained in Raman spectra. Whereas, using the CMO, in the grown C_{60} crystals in a refrigerator for one week, the remarkable results were also not observed.

The results mentioned above are summarized as a graph of figure 5. The open and closed symbols are of the specimens before and after the FEL irradiation, respectively. The intensity ratio was calculated by dividing the intensity of the Ag(2)-derived mode by the Ag(2) mode at 1469 cm^{-1}. In the calculation, the linear baseline was subtracted in advance. Sicne the intensity of the Ag(2) derived mode before the FEL irradiation as shown in figure 1(a) and 4(a) was so small, the plots are not described in figure 5. As shown in figure 3, the specimen after the FEL irradiation did not show the pristine Ag(2) mode at 1469 cm^{-1}, therefore the value is defined as five. In the figure 5, a plot at the left-top area means that the polymerization degree is high. For easy understanding, the transition by the FEL irradiation was described by the dashed arrows in the figure. Along the dashed arrows, the intensity of the Ag(2)-derived mode increased and the peak position downshifted, revealed that the polymerization degree was promoted, in particular the specimen pressed at 7GPa with I_2 doping, see triangles. We found that the I_2 hole doping and the introduction of the rather high pressure were important roles to obtain high polymerized degree. The appearence of the Ag(2)-derived mode aroud 1456 cm^{-1} indicated that the F (fcc structure, space group : Fm3m) phase with the lattice constant of approximately 13.6Å. In addition, the polymerization degree was higher than that of traditional photopolymerization, represented by the peak at 1460 cm^{-1} indicated in the figure as dashed line and notation, even though the polymerization process in this study was photon, the FEL, irradiation [14].

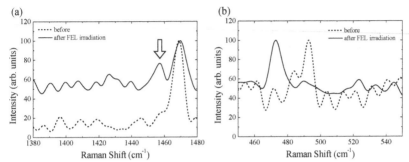

Figure 4. The Raman spectra around (a) Ag(2) mode and (b) Ag(1) mode. The specimen was grown with the saturated CMO and the C_{60} saturation o-xylene solution. As-grown C_{60} crystals showed only Ag(2) mode at 1469 cm^{-1}, and the additional peak at 1457 cm^{-1} appeared after the FEL irradiation. The Ag(1) peak transferred from 493 cm^{-1} to 473 cm^{-1} by the FEL irradiation.

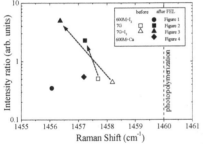

Figure 5. The Raman shift dependence of the intensity ratio, which was calculated by dividing the intensity of the Ag(2)-derived mode by the Ag(2) mode. The dashed arrows means the transition by the FEL irradiation.

SUMMARY

Hole or electron doped C_{60} crystals were grown by the LLIP method to exclude the degradation by oxygen. In the LLIP process, the I_2 dissolved or the CMO saturated BTA were used for hole and electron doping. In all specimens the polymerization degree was promoted, demonstrated by the intensity increasing and the downshift of the position of the Ag(2)-derived mode. In particular, the specimen grown with the I_2 dissolved BTA and the C_{60} saturation o-xylene solution in a refrigerator, pressed at 7GPa, showed the Ag(2)-derived mode at 1456 cm^{-1} without the Ag(2) mode of pristine C_{60} at 1469 cm^{-1}. The position of 1456 cm^{-1} indicated the presence of the F phase, which is polymerized state higher than that of typical photopolymerization.

ACKNOWLEDGMENTS

This work was supported by a Grant-in-Aid (No.20360145) for Scientific Research (B), Japan. The authors are grateful to Laboratory for Electron Beam Research and Application (LEBRA) facility, the CST Nihon Univ., Japan for the use of the FEL.

REFERENCES

1. V. D. Blank, S. G. Buga, G. A. Dubitsky, N. R. Serebryanaya, M. Y. Popov, and B. Sundqvist, *Carbon* **36**, 319 (1998).
2. S. Yamanaka, A. Kubo, K. Inumaru, K. Komaguchi, N. S. Kini, T. Inoue, and T. Trifune, *Phys. Rev. Lett.* **96**, 076602 (2006).
3. D. H. Chi, Y. Iwasa, T. Takano, T. Watanuki, Y. Ohishi, and S. Yamanaka, *Phys. Rev.* **B 68**, 153402 (2003).
4. V. Bkank, M. Popov, G. Pivovarov, N. Lvova, K. Gogolinsky, and V. Reshetov, *Dia. Rel. Mater.* **7**, 427 (1998).
5. Y. Hayakawa, I. Sato, K. Hayakawa, T. Tanaka, K. Yokoyama, K. Kanno, T. Sakai, K. Ishiwata, K. Nakao, and E. Hashimoto, *Nucl. Instrum. Methods Phys. Res.*, Sect. A **507**,404 (2003).
6. H.Nakayama, H. Katayama-Yoshida, *Jpn. J. Appl. Phys.* **41**, L817 (2002).
7. R. Nouchi, K. Masunari, T. Ohta, T. Kubozono, and Y. Iwasa, *Phys. Rev. Lett.* **97**, 196101 (2006).
8. Y. Nakamura, F. Kagawa, K. Kasai, Y. Mera, K. Maeda, *Surf. Sci.* **528**, 151 (2003).
9. Y. Nakamura, Y. Mera, and K. Maeda, *Surf. Sci.* **497**, 166 (2002).
10. H. Yamamoto, and N. Iwata, *Trans. Mater. Res. Soc. Jpn.* **35**, 461 (2010).
11. S. Ando, R. Nokariya, R. Koyaizu, N. Iwata and H. Yamamoto, *Trans. Mater. Res. Soc. Jpn.* **32**, 1251 (2007).
12. N. Iwata, Y. Iio, S. Ando, R. Nokariya, and H. Yamamoto, Mater. Res. Soc. 2007 Fall Proc. 1057-II05-48.
13. K. Miyazawa, and K. Hamamoto, *Materials Research Society* **17**, 2205 (2002).
14. P. C. Eklund, A. M. Rao (eds.), *Fullerene Polymers and Fullerene Polymer Composites* (Springer, 1999), p.169-178.

Mater. Res. Soc. Symp. Proc. Vol. 1284 © 2011 Materials Research Society
DOI: 10.1557/opl.2011.643

Comparison of Epitaxial Graphene on Si-face and C-face 6H-SiC

Shin Mou[1], J. J. Boeckl[1], L. Grazulis[1], B. Claflin[2], Weijie Lu[3], J. H. Park[1], and W. C. Mitchel[1]

[1]Air Force Research Laboratory, Materials and Manufacturing Directorate, Wright-Patterson AFB, OH 45433, USA

[2]Wright State University, Semiconductor Research Center, Dayton, OH 45435, USA

[3]Fisk University, Department of Chemistry, Nashville, TN 37208, USA

ABSTRACT

We present atomic force microscopy (AFM), Hall-effect measurement, and Raman spectroscopy results from graphene films on 6H-SiC (0001) and (000-1) faces (Si-face and C-face, respectively) produced by radiative heating in a high vacuum furnace chamber through thermal decomposition. We observe that the formation of graphene on the two faces of SiC is different in terms of the surface morphology, graphene thickness, Hall mobility, and Raman spectra. In general, graphene films on the SiC C-face are thicker with higher mobilities than those grown on the Si-face.

INTRODUCTION

Graphene, which consists of a few layers of atomic carbon sheets, has recently attracted great interests thanks to its unique electronic and optical properties such as very high mobility and optical transparency. To produce graphene, there are several methods including mechanical exfoliation discovered by Novoselov and Geim et al. in 2004 [1], which allows for investigation of the electronic properties of graphene. Due to the limited size of graphene flakes generated by this method (maximum size about 1 mm^2), other methods have been adapted for applications which require wafer-size graphene films such as epitaxial graphene grown by SiC sublimation [2, 3], metal-catalyst CVD growth [4, 5], and direct carbon deposition [6, 7]. Among these, epitaxial graphene has been proven to produce high quality and highly uniform graphene across the SiC substrate and is also ready for photolithography fabrication with no physical transfer required. Therefore, it is currently the preferred material for high-speed and high-performance graphene field effect transistor (GFET) [8, 9].

It was long known in the SiC community that annealing SiC at high temperatures resulted in the sublimation of Si and the formation of a graphite-like carbon layer on the SiC surface [10], but it was not until Berger et al. [2] conducted thermal decomposition of SiC in ultra high vacuum (UHV) to produce ultra-thin graphite (graphene) on the SiC Si-face that it became an important electronic material. Emtsev et al. [3] later demonstrated high quality growth of the Si-face graphene by decomposition of SiC in atmospheric pressure argon ambient. However, despite the good uniformity of the thickness and the atomic flatness of graphene on the SiC Si-face, the reported Hall mobilities of the ungated graphene are on the order of 1000 cm^2/Vs [11] – lower than those of exfoliated graphene. On the other hand, graphene grown on the SiC C-face has been reported with Hall mobility values of more than 10,000 cm^2/Vs [11]. Nonetheless, thickness uniformity, film stacking, and surface morphology remain issues [12] with the observation that micro van der Pauw (vdP) mobilities range from 100 to 10,000 cm^2/Vs in a recently published article [13]. Motivated by this dramatic difference of graphene films on the

SiC Si- and C-face, we conduct a comparison study of graphene on the two faces of SiC to look for insight into the growth mechanisms.

EXPERIMENT

On-axis CMP 6H-SiC wafers were obtained from II-VI Inc. and normally were diced into 1×1 cm^2 pieces for graphene growth. The annealing process was performed in a graphite element resistive heater furnace (Oxy-Gon Industrial Inc.) in atmospheric pressure Ar ambient at temperatures ranging from 1500°C to 1700°C. No hydrogen etch was applied to the samples beforehand. The annealing times ranged from a few minutes up to 20 minutes. After an oxide removal step in diluted hydrofluoric acid (~15%), samples were immediately placed at the center of the graphite heat zone in the furnace. After pumping down the chamber, Ar gas was purged into the chamber to maintain atmospheric pressure and the temperature was raised to the setpoint target to stay constant for a period of time before cooling down. After the growth, samples went through a series of characterization including AFM (NanoScope IIIa AFM in tapping mode), Raman spectroscopy (Renishaw inVia Raman microscope), and electrical characterization (i.e., two point voltage-current measurements and Hall-effect measurements with four contacts in vdP configurations) to confirm the existence of graphene. Then, graphene samples underwent UV lithography to fabricate Greek cross micro vdP structures for Hall measurements where oxygen plasma etching was used for isolation and Ti/Au was used for contact metal.

DISCUSSION

Graphene films on the SiC Si- and C-face were grown under various conditions (i.e., various temperatures and growth times.) and characterized. AFM surface morphology, Hall measurement data, and Raman spectroscopy will be discussed and compared below under the respective subheadings.

AFM surface morphology

The results from AFM scanning show distinct surface morphologies of graphene on each face of SiC. In Fig. 1, we show AFM morphologies of graphene on both faces grown at various temperatures with growth time of 20 min. At 1500°C, there is step bunching on both the SiC C- and Si-face (Fig. 1 (a) and 1 (d)). Using 2-point electrical measurement (using two probes each with ~ 50 μm tip diameter) and Raman spectroscopy, we conclude that there is no graphene on the SiC Si-face. On the SiC C-face, electrical characterization does not show any sign of conduction. However, we detect graphene with Raman spectroscopy and also find evidence of local graphene formation with wrinkles in the box area around the center of Fig. 1 (a), which is enlarged in the inset. Most likely, graphene on the C-face nucleates locally in some regions at 1500°C but does not cover the whole surface continuously. Therefore, we believe that graphene on the C-face nucleates at lower temperature than on the Si-face and the formation mechanism is not the same as that on the Si-face, where graphene systematically nucleates from step edges and grows layer by layer [14]. Our observation indicates nucleation occurs at random locations, which is consistent with the hypothesis that the C-face graphene nucleates from defects [15]. When the growth temperature is raised to 1600°C, the appearance of the Si-face (Fig. 1 (e)) looks similar to that of samples grown at 1500°C [16], but graphene has formed throughout the surface based on electrical characterization and Raman spectroscopy. The smooth surface indicates that the topology of graphene follows the morphology of the underlying SiC wafer on the Si-face. For the C-face graphene, the surface morphology looks completely different than the morphology seen for C-face growth at1500°C. To be more specific, the atomic step bunching originating from the SiC is washed out and replaced by random raised and lowered areas with nets of wrinkles spreading across the whole field of view (Fig. 1 (b)). The reason is likely that due to various local nucleation rates and threshold temperatures, graphene grows faster in some areas than in the others. The lowered regions represent areas where graphene is thicker [12] because more SiC bilayers have decomposed into graphene. The wrinkles likely result from the lattice mismatch between graphene and the SiC C-face. On the other hand, the bond between graphene and the Si-face of SiC is relatively strong, and therefore the Si-face graphene has a strain level about 1% [17] and remains wrinkleless when it is only a-few-atomic-layer thick. When the growth temperature is increased to 1700°C, the surface morphologies for both faces (Fig. 1 (c) for C-face and Fig. 1 (f) for Si-face) remain similar to those grown at 1600°C, respectively. For C-face graphene, the height between raised and lowered regions becomes larger and the numbers of wrinkles increase. These are indications that the graphene film is thicker. Regarding the Si-face graphene, the surface step bunching remains crisp; the surface roughness and the surface morphology do not seem to change too much compared to growth at1600°C [18].

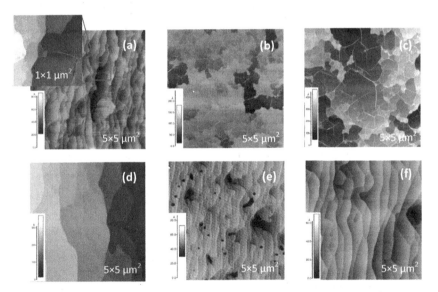

Fig. 1. AFM images of annealed 6H-SiC surfaces for annealing conditions of (a) 1500°C, 20 min on the C-face, (b) 1600°C, 20 min on the C-face, (c) 1700°C, 20 min on the C-face, (d) 1500°C, 20 min on the Si-face, (e) 1600°C, 20 min on the Si-face, and (f) 1700°C, 20 min on the Si-face. The inset of Fig. 1 (a) shows and enlarges the box area.

Hall measurement

Similar to the AFM surface morphology, the Hall measurement results are distinct between C-face and Si-face graphene. We conduct Hall measurements on two different kinds of test structures. Bulk measurement, which is often done right after the growth, uses four contacts on the four corners of the 1×1 cm^2 pieces to conduct the Hall-effect measurement in a vdP configuration. We also microfabricate Greek cross vdP structures with different sizes (width from 2 μm to 250 μm) on graphene samples and perform Hall measurements. The results are summarized in Fig. 2 (a) and Fig. 2 (b) in scatter plots of Hall mobility versus sheet carrier density and sheet resistivity versus sheet carrier density, respectively. For the Si-face graphene, we find that the correlation between mobilities and carrier densities of the Si-face graphene exhibit power-law behavior (nearly straight line on a log-log plot). Moreover, the correlation is similar for bulk or micro vdP measurements. Please note, all the micro vdP data presented here for the Si-face graphene comes from a single sample and the total 12 devices have a tight distribution of mobility from 378 to 490 cm^2/Vs (low variability). On the other hand, for C-face samples, even though the bulk measurement and micro vdP data fall on lines in Fig. 2, respectively, these two lines do not overlap with each other as opposed to results reported previously [11]. Please note, although our C-face SiC surface is well polished to a degree that we do not observe scratches with AFM, the average surface roughness is still larger than that of the Si-face SiC and we did not perform any pre-growth hydrogen etch. Potentially this increased roughness can induce more growth inhomogeneity and result in the discrepancy seen between the Hall data of macro bulk samples and micro vdP structures. More studies need to be done to understand the root cause of this discrepancy. Comparing the vdP Hall data of the C-face and the Si-face graphene, the maximum mobility on the C-face is higher (~2300 cm^2/Vs). Similar to the Si-face measurements, all the vdP Hall data on the C-face graphene comes from one sample. The vdP mobilities of the C-face graphene have a much larger variability (values from 200 to 2300 cm^2/Vs) than that of the Si-face. At this point, we cannot confirm whether the variability is a result of material inhomogeneity, fabrication process variation, starting SiC surface, or something else. In summary, the C-face graphene has a higher maximum mobility value but also a higher variability in Hall data and we need further studies to understand the root cause of this variability.

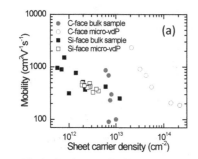 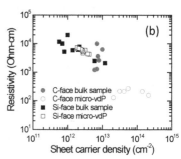

Fig. 2. Graphene Hall effect data of (a) mobility versus sheet carrier density and (b) sheet resistivity versus sheet carrier density. 1×1 cm^2 bulk graphene samples on the Si-face graphene are represented in black filled squares, micro vdPs on the Si-face graphene in black hollow squares, 1×1 cm^2 bulk samples on the C-face graphene in red filled circles, and micro vdPs on the C-face graphene in red hollow circles.

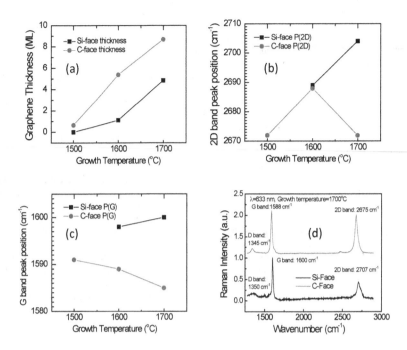

Fig. 3. Growth temperature versus (a) estimated graphene thickness, (b) the 2D-band peak position, (c) the G-band peak position. (d) Representative Raman spectra of the Si-face graphene and the C-face graphene grown at 1700°C.

Raman spectroscopy

Raman spectroscopy measurements have been performed on epitaxial graphene [19-21] and reveal important information about thickness, charge carrier density, disorder, and strain. The main features in the Raman spectra of carbon-based materials are the so-called G and D peaks, which lie at around 1560 and 1360 cm^{-1} respectively for visible excitation (633 nm laser in this work) and an overtone peak around 2700 cm^{-1}, the so called the 2D or G' peak also present in graphitic structures. In this work, for every piece of sample, our methodology is that we take average several Raman spectra at different locations. Then, we use the attenuation of the SiC substrate Raman intensity to estimate the thickness of graphene as described in the work by Shivaraman et al. [19] and the resulting thickness is plotted in Fig. 3 (a). Qualitatively, the estimated thickness agrees with the AFM observation on the C-face graphene. For 1500°C growth, the average graphene thickness on the SiC C-face is estimated to be 0.7 monolayer

(ML), which is consistent with the observation from the previous section that the C-face of SiC is only partially covered by graphene. On the other hand, due to the absence of the 2D peak in Raman spectra for Si-face graphene grown at 1500°C, we conclude that there is no graphene present there. For 1600°C growth, the estimated C-face graphene thickness increases to 5.4 ML while it is only 1.1 ML on the Si-face of SiC (thinner due to the higher nucleation temperature and slower growth rate.) For growth at 1700°C, graphene thickness increases to 8.7 ML and 5.5 ML on the C-face and the Si-face of SiC, respectively. This confirms the observation from the AFM morphology that graphene on the C-face SiC nucleates at a lower temperature and grows more rapidly than graphene on the Si-face. Another distinction is that the G and 2D bands of the Si-face graphene shift to higher wavenumber (a red shift) compared to those of the C-face graphene (see Fig. 3 (b), (c), and (d).) There are several potential causes including different doping levels and strain [20]. While doping level differences could produce a substantial shift, it should give rise to a blue shift instead of the red shift for the G band based on the measured Hall carrier density [21] (the C-face graphene has larger doping/carrier density than that of the Si-face graphene by an order of magnitude.) Strain is likely a cause for the G band red shift. The G band of the C-face graphene has a peak position around 1587 cm^{-1}, which is consistent with the G-band peak position of unstrained exfoliated monolayer graphene at same laser excitation wavelength (633 nm) [22]. At the same time, the red shift of the Si-face graphene is on the order of 10 cm^{-1}, which is comparable to the red shift attributed to strain found by Ni et al. [20]. It is generally believed that the Si-face graphene is attached to the 6R3 buffer layer and therefore has a higher level of strain [17, 10]. On the other hand, the C-face graphene simulates exfoliated graphene and has less strain and more wrinkles. This potentially results in higher mobilities (than the Si-face graphene.)

CONCLUSIONS

We have carried out a comprehensive study using AFM, Hall measurement, and Raman spectroscopy to compare graphene grown on the Si- and C-faces of SiC at various temperatures. While the C-face graphene possesses larger surface roughness resulting from steps and wrinkles, it has higher mobility possibly due to lower strain confirmed by Raman spectroscopy. However, the C-face graphene appears to have larger variability in the Hall mobilities and it tends to grow faster and thicker. Further investigations will be conducted to determine if this variability is due to material inhomogeneity, device microfabrication issues, or SiC pre-growth surface conditions.

ACKNOWLEDGMENTS

This work is supported by AFOSR (Dr. Harold Weinstock) and Air Force Research Laboratory (Dr. Katie Thorp and Dr. Augustine Urbas.) The authors wish to thank Mr. Gerald Landis, Mr. John Hoelscher, and Mr. Timothy Cooper for technical assistance. We would also like to thank Dr. D. Tomich for discussion concerning the Raman spectroscopy.

REFERENCES

1. K. S. Novoselov, A. K. Geim, S. V. Morozov, D. Jiang, Y. Zhang, S.V. Dubonos, I. V. Grigorieva, and A.A. Firov, *Science* **306**, 666 (2004).
2. C. Berger, Z. Song, X. Li, X. Wu, N. Brown, C. Naud, D. Mayou, T. Li, J. Hass, A. N. Marchenkov, E. H. Conrad, P. N. First, and W. A. de Heer, *Science* **312**, 1191 (2006).
3. K. V. Emtsev, A. Bostwick, K. Horn, J. Jobst, G. L. Kellog, L. Ley, J. L. McChesney, T. Ohta, S. A. Reshanov, J. Röhrl, E. Rotenberg, A. K. Schmid, D. Waldmann, H. B. Weber, and Th. Seyller, *Naturer Mater.* **8**, 203 (2009).
4. X. Li, Y. Zhu, W. Cai, J. An, S. Kim, J. Nah, D. Yang, R. Piner, A. Velamakanni, I. Jung, E. Tutuc, S. K. Banerjee, L. Colombo and R. S. Ruoff, *Science* **324**, 1312 (2009).
5. K. S. Kim, Y. Zhao, H. Jang, S. Y. Lee, J. M. Kim, K. S. Kim, J.-H. Ahn, P. Kim, J.-Y. Choi, and B. H. Kim, *Nature* **457**, 706 (2009).
6. J. Park, W. C. Mitchel, L. Grazulis, H. E. Smith, K. G. Eyink, J. J. Boeckl, D. H. Tomich, S. D. Pacley, and J. E. Hoelscher, *Adv. Mater.* **22**, 4140 (2010).
7. J. Hwang, V. B. Shields, C. I. Thomas, S. Shivaraman, D. Hao, M. Kim, A. R. Woll, G. S. Tompa, and M. G. Spencer, *J. Cryst. Growth* **312**, 3219 (2010).
8. Y.-M. Lin, C. Dimitrakopoulos, K. A. Jenkins, D. B. Farmer, H.-Y. Chiu, A. Grill, Ph. Avouris, *Science* **327**, 662 (2010).
9. J. S. Moon, D. Curtis, S. Bui, M. Hu, D. K. Gaskill, J. L. Tedesco, P. Asbeck, G. G. Jernigan, B. L. VanMil, R. L. Myers-Ward, C. R. Eddy, Jr., P. M. Campbell, and X. Weng, *IEEE Electron Device Lett.* **31**. 260 (2010).
10. A. J. van Bommel, J. E. Crombeen, and A. van Tooren, *Surf. Sci.* **48**, 463 (1975).
11. J. L. Tedesco, B. L. VanMil, R. L. Myers-Ward, J. M. McCrate, S. A. Kitt, P. M. Campbell,2 G. G. Jernigan, J. C. Culbertson, C. R. Eddy, Jr.,1 and D. K. Gaskill, *Appl. Phys. Lett.* **95**, 122102 (2009).
12. Luxmi, P. J. Fisher, N. Srivastava, R. M. Feenstra, Yugang Sun, J. Kedzierski, P. Healey, and Gong Gu, *Appl. Phys. Lett.* **95**, 073101 (2009).
13. J. A. Robinson, M. Wetherington, J. L. Tedesco, P. M. Campbell, X. Weng, J. Stitt, M. A. Fanton, E. Frantz, D. Snyder, B. L. VanMil, G. G. Jernigan, R. L. Myers-Ward, C. R. Eddy, Jr., and D. K. Gaskill, *Nano Lett.* **9**, 2873 (2009).
14. S. W. Poon, W. Chen, E. S. Tok, and Andrew T. S. Wee, *Appl. Phys. Lett.* **92**, 104102 (2008).
15. J. L. Tedesco, G. G. Jernigan, J. C. Culbertson, J. K. Hite, Y. Yang, K. M. Daniels, R. L. Myers-Ward, C. R. Eddy, J. A. Robinson, K. A. Trumbull, M. T. Wetherington, P. M. Campbell, and D. K. Gaskill, *Appl. Phys. Lett.* **96**, 222103 (2010).
16. After the step bunching, the terrace width normally ranges from 0.5 μm to 2 μm. The step heights are higher when the terraces are wider where Fig. 1 (d) is an example.
17. Joshua A. Robinson, Conor P. Puls, Neal E. Staley, Joseph P. Stitt, Mark A. Fanton, Konstantin V. Emtsev, Thomas Seyller, and Ying Liu, *Nano Lett.*, **9**, 964 (2009).
18. Fig. 1 (e) shows pits holes on graphene while Fig. 1 (f) does not. We cannot confirm the origin of the pit holes but it is probably not due to the different growth temperature since they show up randomly in samples at various growth temperatures.
19. S. Shivaraman, M.V.S. Chandrashekhar, J. J. Boeckl, and M. G. Spencer, *J. Electron. Mater.* **38**, 725 (2009).

20. Z. H. Ni, W. Chen, X. F. Fan, J. L. Kuo, T. Yu, A. T. S. Wee, and Z. X. Shen, *Phys. Rev. B* **77**, 115416 (2008).
21. A. Das, S. Pisana, B. Chakraborty, S. Piscanec, S. K. Saha, U. V. Waghmare, K. S. Novoselov, H. R. Krishnamurthy, A. K. Geim, A. C. Ferrari, and A. K. Sood, *Nat. Nanotech.* **3**, 210 (2008).
22. Andrea C. Ferrari, *Solid State Comm.* **143**, 47 (2007).

CNTs Growth, Exploring Novel CNT Growth Techniques and Growth Mechanisms I

Mater. Res. Soc. Symp. Proc. Vol. 1284 © 2011 Materials Research Society
DOI: 10.1557/opl.2011.644

Growth of diameter-modulated single-walled carbon nanotubes through instant temperature modulation in laser-assisted chemical vapor deposition

M. Mahjouri-Samani, Y. S. Zhou, W. Xiong, Y. Gao, M. Mitchell, and Y. F. Lu[*]
Department of Electrical Engineering, University of Nebraska-Lincoln, Lincoln, NE 68588-0511

ABSTRACT

The diameter of individual single-walled carbon nanotubes (SWNTs) was successfully modulated along their axes by instant temperature control in a laser-assisted chemical vapor deposition (LCVD) process. SWNTs were grown using different temperature profiles to investigate the effects of temperature variation on their growth. Due to the inverse relationship between SWNT diameter and growth temperature, SWNTs with ascending diameters were obtained by reducing the LCVD temperature from high to low. The diameter-modulated SWNTs were grown across a pair of Mo electrodes to form field-effect transistors (FETs) for investigation of their electronic transport properties. Fabricated devices demonstrated properties similar to Schottky diodes, implying different bandgap structures at the ends of the SWNTs. Raman spectroscopy, transmission electron microscopy, and electronic transport characteristics were studied to investigate the influence of temperature variation on the structural and electronic characteristics of SWNTs.

INTRODUCTION

Remarkable electronic and physical characteristics of single-walled carbon nanotubes (SWNTs) have made them an interesting material for nanoscale electronics and optical devices such as transistors and sensors [1]. One of the most fascinating characteristic of SWNTs is their ability to be either metallic or semiconducting with variable bandgaps depending on their diameters and chiral vectors [2, 3]. However, many researchers treat this as a challenging issue for fabrication of SWNT-based devices with uniform electronic properties. These variations of structures, electronic types, and bandgaps have been limiting the applications of SWNTs in electronic and optical devices. Thus, large amounts of research have been done to control and separate SWNTs with different chiralities and diameters in the pre and post synthesis processes [4-8]. Tremendous efforts and focus on the uniformity issue have prevented researchers from taking these variations as an advantage. Recently, diameter variation in the growth of ultralong SWNTs by altering the growth temperature has been reported [9]. However, the growth and characterization of diameter-modulated SWNTs have not been investigated for their applications in nanoscale devices.

In this study, we took these variations in diameters and bandgaps as an advantage and were able to alter their diameters during the growth in a laser-assisted chemical vapor deposition (LCVD) process. SWNTs with continuous change of diameters along their length were grown, implying the possibility to form array of straddling bandgaps along the individual semiconducting SWNTs. Equation (1) clearly describes the inverse proportionality of the bandgaps of the semiconducting SWNTs to their diameters [10].

---

[*] Correspondence should be addressed to Prof. Y. F. Lu, Tel: (402) 472-8323, Fax: (402) 472-4732, Email: ylu2@unl.edu.

$$E_g = \frac{(2\gamma_o \times a_{C-C})}{d_t}$$ (1)

γ_o is the C–C tight-binding overlap energy = ~2.7 eV, a_{C-C} = 0.142 nm, the nearest-neighbor C-C distance. Also, d_t is the diameter of an SWNT in terms of the chiral vector indices, n and m, which is described as

$$d_t = \frac{a}{\pi} \sqrt{(n^2 + m^2 + nm)}.$$ (2)

It is also reported that the diameter of the tubes is inversely proportional to the growth temperature during the SWNT growth process [8].Therefore, tuning of the diameters of SWNTs can potentially modulate the bandgaps of semiconducting SWNTs. New electronic and optical properties can be expected from the diameter-modulated SWNTs such as a diode-like *I-V* performance and wide range of optical absorption.

EXPERIMENTAL SECTION

Vertically aligned carbon-nanotube (CNT) forests and individual CNTs bridging predefined Mo electrodes on Si/SiO$_2$ substrates were prepared using an LCVD process. P-type silicon wafers coated with a 2-μm thick oxide layer were used as the substrates. Tri-layer Al/Fe/Al catalyst film (30 nm Al, 0.8 nm Fe and 1 nm Al separately) was sputtered on the substrates to grow vertically aligned CNTs. Using conventional photolithography, 200 nm thick Mo patterns covered with the tri-layer catalyst film were fabricated and used as the electrodes for fabricating SWNT bridges. A vacuum chamber was used to grow CNTs with a background pressure of 1×10^{-3} Torr. The catalyst film was first annealed at 500 °C for 1 min in an ammonia atmosphere of 10 Torr to make catalyst nanoparticles. Acetylene and ammonia gases with a volume ratio of 1:10 and pressure of 10 Torr were then introduced into the chamber. A continuous wave CO$_2$ laser (Synrad, Firestar v40, wavelength of 10.6 μm) was used as a heat source. The substrate temperature was controlled instantaneously by adjusting the incident laser power. A non-contact infrared pyrometer (Omega, OS3750) was used to monitor the substrate temperature. A field-emission transmission electron microscope (FEI Tecnai G2 F30) was used to observe the diameter changes along the CNTs. A field-emission scanning electron microscope (FE-SEM, Hitachi S4700) was used to capture scanning electron micrographs of the CNTs. The electronic transport characterization was carried out using an Agilent HP4155C semiconductor parameter analyzer. Raman spectra were acquired using a Renishaw inVia dispersive micro-Raman spectrometer equipped with a standard microscope objective (×100), providing a focal spot with a diameter of ~ 1 μm.785, 514 and 633 nm lasers were used as the excitation sources for Raman spectroscopy.

RESULTS AND DISCUSSION

According to our results, high-quality CNTs can be obtained in a temperature range from 450 to 650 °C through LCVD. Therefore, these two temperatures were chosen to be the low and high temperature limits in this study. Multi-walled CNTs (MWNTs) were dominantly grown at 450 °C while SWNTs mostly grew when the temperature was kept at 650 °C. As shown in Fig.

1(b), Raman spectrum of the CNTs grown at 450 °C show no radial breading mode (RBM) peak and the G-band is located very close to the graphite peak (1582 cm^{-1}) indicating the growth of MWNTs. However, RBM peaks with right shifted G-band were observed (Fig 1(a)) from the CNTs grown at 650 °C indicating the existence of SWNTs. Due to our interest in the electronic properties of the SWNTs, growth process was started from high temperature to avoid dominant growth of MWNTs. Therefore, CNT growth with descending temperature profiles starting from 650 °C was investigated due to the dominant nucleation of the SWNTs at this temperature.

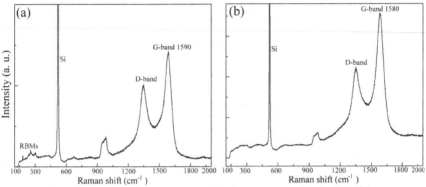

Figure 1. (a) and (b) show the Raman spectra of the CNTs grown at 650 and 450 °C respectively.

SWNTs with increasing diameters were grown by reducing the growth temperature from 650 to 450 °C. The temperature was decreased from $T_{initial} = 650$ to $T_{final} = 450$ °C in 10 steps of 20 °C, maintaining $\Delta t = \left| T_{initial} - T_{final} \right| = 200$ °C. The process was also carried out in different reaction time intervals, while maintaining $\Delta t = 200$ °C, to investigate the influence of the rate of temperature change on CNT diameter variation. Vertically aligned SWNTs were grown on Si/SiO$_2$ substrates and subsequently characterized by Raman spectroscopy and TEM imaging in different regions of the SWNTs. Figure 2 shows the TEM images of a typical SWNT with an increasing diameter from one end to another. Unfortunately, due to the background noise coming from the carbon film supporting the SWNTs on the TEM grids and the agglomeration of SWNTs, it was challenging to obtain high quality TEM images. The RBM peaks distribution of the vertically aligned SWNTs was obtained and studied using 514 and 633 nm laser excitations as shown in Fig. 3(c). RBM peaks at low and high frequencies were present at the starting end of the SWNTs proving the growth of SWNTs at our high growth temperature with random diameters. However, on the other side of the tubes, a red-shifted broad RBM is observed. The broadband RBM peak is possibly because of an overlap between the red-shifted RBM due to increase in diameters. The spectrum obtained with 514 nm from the top of the samples is very noisy and no information could be concluded from.

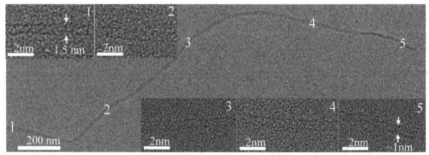

Figure 2. TEM image of a 1 μm long SWNT grown with a descending temperature profile. The close up views, (1–5), corresponding to the labeled areas of the tube, show about 0.4 nm variation in the diameter from one end to the other.

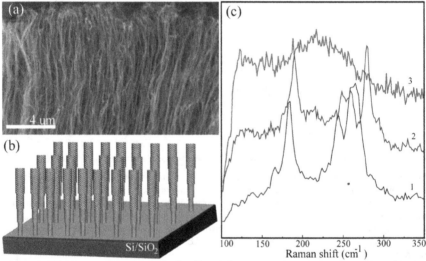

Figure 3. SEM image shows the vertically aligned diameter-modulated CNTs (a) followed by its illustrative schematic (b). Raman spectra 1 and 2 in (c) are obtained from the bottom of the sample with 514 and 633 nm laser excitations respectively. Spectrum number 3 is obtained from the top of the sample with 633 nm excitation. No peak was observed using the 514 nm excitation.

Since the TEM observation of the diameter change in a long tube is not clearly visible, diameter-alternating CNTs were grown and observed under the TEM as shown in Fig 4. Diameter-alternating CNTs were observed when the growth temperature was periodically alternated by an automatic shutter placed in front of the laser. First, the laser power and shutter open-time were adjusted to set the temperature of the substrate to around 650 °C. Then, the shutter close-time was set to a value where the temperature drops to around 450 °C. It was found that it takes about 0.3 seconds to reduce the temperature from 650 to 450 °C and vice versa. Figures 4 show the diameter modulation for the shutter speed of 0.3 sec open- and 0.3 sec close-time. Experiments are under investigation to see the effect of exposure time and the magnitude of temperature change on the diameter of the tube as well.

Figure 4. TEM images of the diameter-alternating CNTs grown with periodic modulation of the temperature.

Individual diameter-modulated SWNTs were grown across Mo electrodes to investigate their electronic transport characteristics and bandgap variations due to the change in their diameters. *I-V* characteristics and Raman spectra of the devices with both uniform-diameter SWNT (Fig. 5) and diameter-modulated SWNT (Fig. 6) were measured to determine their electronic transport characteristics and diameter modulations.

In agreement with previously reported works, a device with a uniform-diameter semiconducting SWNT has almost symmetrical *I-V* curve, Fig. 5(d), which can be attributed to the similar SWNT/Mo contacts at the two ends of the SWNT. Since, both ends of the uniform-diameter semiconducting SWNT have the same diameter and hence bandgap, two identical but opposite Schottky barriers with the metallic electrodes are formed. However, type of physical contact and interaction of the tube with other tubes on the electrodes might have influence on the *I-V* curves resulting in semi symmetrical behavior. These devises did not show strong diode-like behavior at zero V_g bias. Figure 5(b) shows the Raman spectra of the device with similar RBM peaks at both ends of the tube. The schematic of the bandgap diagram of the device is shown in the inset of Fig. 5(d).

On the other hand, the device with a diameter-modulated semiconducting SWNT forms two different contacts with the metallic electrodes. A Schottky contact is formed between the Mo electrode and one end of the tube with a smaller diameter (having wider bandgap). On the other side, Ohmic contact is formed between the Mo electrode and the other end of the tube with a larger diameter (smaller bandgap). The schematic of the band diagram of the device is shown in the inset of Fig. 6(d). When the device was positively biased ($V_{ds} > 0$) and the Schottky contact (source) was grounded, positive current value was obtained. When negatively biased ($V_{ds} < 0$), no current flowed through the device. SWNT-based Schottky diodes have been fabricated and reported by others using different methods [11, 12]. Schottky or Ohmic contacts can be formed between the SWNT and different electrode materials, depending on the work function of the electrode material to contact with. However, this increases the fabrication complexity, time, and cost. Using a diameter-modulated SWNT grown in this study, different bandgaps at the two ends of the SWNT can form a diode by simply bridging two electrodes of the same metal with a diameter-modulated SWNT (e.g., Mo with a work function of about 4.6 eV). However, it is observed that the magnitude of the current is much smaller in diameter-modulated SWNTs than those in their uniform counterpart, possibly due to the electron scattering caused by the defects created in the process of diameter modulations.

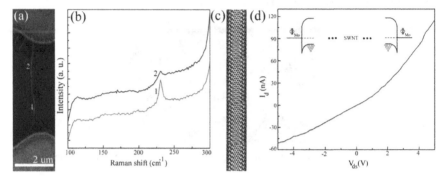

Figure 5. (a) SEM image of a single SWNT across the Mo electrodes; (b) corresponding Raman spectra obtained from the numbered locations in (a) with a constant RBM peak position indicating the uniformity of the SWNT diameter; (c) the schematic representation of the SWNT structure; and (d) nearly symmetrical $I–V$ curve with an inset indicating the expected band diagram of the device.

Likewise, micro-Raman spectroscopy was conducted (using 785 nm excitation laser) on the SWNT of the same device by mapping the SWNT to investigate the diameter modulation and corresponding bandgap variation along the SWNT utilizing the Kataura plot. As shown in Fig. 6(b), RBM peak of the diameter-modulated SWNT shifts from higher frequency (grown with higher temperature) to the lower frequency (grown with lower temperature). Based on the Raman spectra, the Kataura plot, and the bandgap equation provided, corresponding diameter, bandgap, and n, m indices of the SWNT can be calculated. However, it should be noted that it is possible to have intermediate RBM frequencies, either metallic or semiconducting, that are not in resonance with the laser excitation source.

CONCLUSIONS

SWNTs with diameter modulation along their tube axes can be grown by precisely controlling the substrate temperature in a laser-assisted CVD process. A CO_2 laser used as the heat source can quickly vary the temperature by adjusting the laser power. The modulation of the SWNTs' diameter can induce a bandgap change along their lengths. Due to the bandgap modulation, Schottky diodes can be formed with the SWNTs across the metallic electrodes of the same metal such as Mo. The end of the SWNT with the smaller diameter corresponding to wider bandgap forms a Schottky contact with the Mo electrode while the other end with the larger diameter corresponding to the smaller bandgap forms an Ohmic contacts with the Mo electrode. Individual SWNTs with variable bandgaps may have potential applications in photovoltaic devices due to the asymmetric built-in potential at tube/metal contacts and the possibility of absorption of a wide spectrum within a single SWNT.

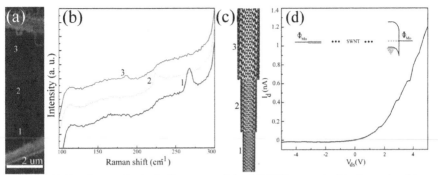

Figure 6. (a) SEM image of a single diameter-modulated SWNT across the Mo electrodes; (b) corresponding Raman spectra obtained from the numbered locations in (a) with shifted RBM peaks indicating the variation of the SWNT diameter; (c) the schematic representation of the SWNT structure; and (d) asymmetrical diode-like *I–V* curves with an inset indicating the expected band diagram of the device.

ACKNOWLEDGMENTS

This research work was financially supported by National Science Foundation (ECCS 0621899 and ECCS 0652905) and Nebraska Research Initiative. We are grateful to Drs. You Zhou, Natale Joseph Ianno and Lanping Yue for technical support.

REFERENCES

1. R. H. Baughman, A. A. Zakhidov, W. A. de Heer, *Science* **297,** 787-792 (2002).
2. T. W. Odom, J. L. Huang, P. Kim, C. M. Lieber, *Nature* **391,** 62-64 (1998).
3. S. M. Bachilo, M. S. Strano, C. Kittrell, R. H. Hauge, R. E. Smalley, R. B. Weisman, *Science* **298,** 2361-2366 (2002).
4. K. Maehashi, Y. Ohno, K. Inoue, K. Matsumoto, *Appl. Phys. Lett.* **85,** 858-860 (2004).
5. S. Reich, L. Li, J. Robertson, *Chem. Phys. Lett.* **421,** 469-472 (2006).
6. M. Mahjouri-Samani, Y. S. Zhou, W. Xiong, Y. Gao, M. Mitchell, Y. F. Lu, *Nanotechnology* **20,** 495202 (2009).
7. M. S. Arnold, A. A. Green, J. F. Hulvat, S. I. Stupp, M. C. Hersam, *Nature nanotechnology* **1,** 60-65 (2006).
8. W. Song, C. Jeon, Y. S. Kim, Y. T. Kwon, D. S. Jung, S. W. Jang, W. C. Choi, J. S. Park, R. Saito, C. Y. Park, *ACS Nano* **4,** 1012-1018 (2010).
9. Y. Yao, X. Dai, R. Liu, J. Zhang, Z. Liu, *J. Phys. Chem. C.* **113,** 13051-13059 (2009).
10. J. W. G. Wildo, L. C. Venema, A. G. Rinzler, R. E. Smalley, C. Dekker, *Nature* **391,** 59-62 (1998).
11. H. M. Manohara, E. W. Wong, E. Schlecht, B. D. Hunt, P. H. Siegel, *Nano Lett.* **5,** 1469-1474 (2005).
12. C. Lu, L. An, Q. Fu, J. Liua, H. Zhang, J. Murduck, *Appl. Phys. Lett.* **88,** 133501 (**2006**).

Mater. Res. Soc. Symp. Proc. Vol. 1284 © 2011 Materials Research Society
DOI: 10.1557/opl.2011.645

A CMOS Compatible Carbon Nanotube Growth Approach

Daire Cott[1], Masahito Sugiura[2], Nicolo Chiodarelli[1,3],Kai Arstila[1], Philipe M. Vereecken [1,4],
Bart Vereecke[1], Sven Van Elshocht[1] , and Stefan De Gendt[1,5];
[1]IMEC, 75 Kapeldreef, Leuven, Belgium
[2]Tokyo Electron Ltd., Technology Development Center, 650 Mitsuzawa, Hosaka-cho, Nirasaki,
Yamanashi 407-0192, Japan
[3]Electrical Engineering, Katholieke Universiteit Leuven, Leuven, Belgium;
[4]Center for Surface Chemistry and Catalysis, Katholieke Universiteit Leuven, Leuven, Belgium;
[5] Department.of Chemistry, Katholieke Universiteit Leuven, Leuven, Belgium.

ABSTRACT

In future technology nodes, 22nm and below, carbon nanotubes (CNTs) may provide a viable alternative to Cu as an interconnect material. CNTs exhibit a current carrying capacity (up to 10^9 A/cm^2), whilst also providing a significantly higher thermal conductivity (SWCNT ~ 5000 WmK) over Copper (10^6 A/cm^2 and ~400WmK). However, exploiting such properties of CNTs in small vias is a challenging endeavor. In reality, to outperform Cu in terms of a reduction in via resistance alone, densities in the order of 10^{13} CNTs/cm^2 are required. At present, conventional thermal CVD of carbon nanotubes is carried out at temperatures far in excess of CMOS temperature limits (400 °C). Furthermore, high density CNT bundles are most commonly grown on insulating supports such as Al$_2$O$_3$ and SiO$_2$ as they can effectively stabilize metallic nanoparticles at elevated temperatures but this limits their application in electronic devices. To circumvent these obstacles we employ a remote microwave plasma to grow high density CNTs at a temperature of 400 °C on conductive underlayers such as TiN. We identify some critical factors important for high-quality CNTs at low temperatures such as control over the catalyst to underlayer interaction and plasma growth environment while presenting a fully CMOS compatible carbon nanotube synthesis approach

INTRODUCTION

The electronic properties, conducting/semiconducting of carbon nanotubes (CNTs) are determined by their molecular structure in turn determining tube diameter, as a result each single walled CNT can be considered as a macromolecule with distinct chirality. For applications such as high volume field effect transistors (FET) fabrication strict control over their semiconducting properties is a prerequisite. Thus, for transistors, CNT synthesis can be considered as selective chiral molecular synthesis and at present the synthesis of one individual chiral controlled entity by CVD is a challenge yet to be realized. In the near future a more attainable goal for CNTs in microelectronics may be their use for interconnections as they could provide a viable replacement to Cu and W at sub 22nm dimensions. Here the chirality effect can be somewhat overlooked and the metallic nature [1], density (number of CNTs/cm^2) [2], and synthesis compatibility with current CMOS processing become the critical factors [3]. For carbon nanotubes as on-chip vertical interconnections one integration target is at the metal contact level directly above the active transistor device as it will become increasingly challenging to completely fill small diameter (~20nm) high aspect ratio with vias with Cu and W [4]. Ideally, to compete with Cu in terms of resistance alone a CNT density 10^{13} CNTs/cm^2 would be required.

Indeed, one MWCNT with 10 walls filling an 11 nm contact hole would satisfy this density requirement.

At present the most scalable technique to grow high density CNTs within a semiconductor platform is chemical vapour deposition (CVD) of low molecular weight gaseous hydrocarbons such as CH_4 and C_2H_2 catalyzed by an annealed thin (1-2nm) metal film, typically Ni, Co or Fe. In CVD the energy required to break down the reactant hydrocarbon precursor into graphene sheets comes solely from the thermal energy supplied to the metal catalyst particle and its surrounds. For MWCNT growth with larger diameters the rate limiting step is comparable to the diffusion of carbon through the bulk catalyst [5] typically of the order of 1.5eV thus limiting growth at CMOS compatible temperatures, being typically below 400 °C. To reduce this energy barrier i.e. temperature, the CVD environment can be RF (13.56MHz) or microwave (2.45GHz) plasma enhanced. However, such plasma environments most often than not produce significant ion-bombardment that tends to etch small diameter CNTs i.e. single walled, thus favouring the growth of more disordered large diameter carbon nanostructures such a carbon nanofibres (CNFs) [6] unsuitable to meet the density requirements for interconnections. To circumvent this one solution is to have the plasma source positioned remotely with respect to the targeted growth area to maximise the amount of active CNT growth species, control etchant concentrations, whilst minimizing ion-bombardment reaching the catalyst [7].

Employing (PE)CVD, probably the most important point and one sometimes overlooked when considering CNTs as interconnections is the underlying layer on which CNTs are grown from. To effectively exploit the high conductivity of CNTs it is essential to have them directly grown on a conductive underlayer. Depending on the technology node and level of interconnect these conductive underlayers will most likely be metal silicides ($CoSi_2$, NiSi or NiPtSi), barrier layers (TiN and TaN), or directly on silicon. The supporting underlayer in combination with the type and thickness of metal catalyst film plays a crucial role in determining the length, diameter and type of carbon structure formed be it single walled, double multi walled (MW), or fiber-like. In the case of oxide based underlayers such as TiO_2, SiO_2 and in particular Al_2O_3 [8] it has been shown that they can support highly active metal particles in well defined size distributions capable of nucleating and catalyzing the growth of SWCNTs up to mm in lengths [9]. However, for via applications the overall conductivity of a via will be dominated by the resistance of such an oxide underlayer [10]. To our knowledge, there are no reports of SWCNT growth directly on bare conducting supports. Recent advances in SWCNT synthesis on Ni-based alloys with Cr or Fe such as inconel have been shown to be effective substrates to grow SWCNTs [11]. However, these metal foils still require an oxide interlayer such as Al_2O_3 to successfully nucleate SWCNTs in high density. As mentioned previously it may be possible to avoid exclusive SWCNT growth and exploit MWCNTs in small via dimensions. Herein, we describe CNT growth at CMOS compatible temperatures (400°C) on Titanium nitride (TiN), commonly used as a liner or metal diffusion barrier in via structures. TiN has an increased resistance to oxidation making it an interesting candidate to support CNT growth and further allow for an effective electrical contact with the as grown CNTs. Similar work [12] has been carried out at higher temperature (~ 800 °C), temperatures far in excess of those imposed by current CMOS processing constraints. We outline the influence of annealing conditions for Ni on TiN and describe the difference between the nanoparticle size and CNT diameter. The learning is applied to the growth of CNT via bundles for potential use as CNT interconnections.

EXPERIMENTAL DETAILS

The blanket catalyst films were made up from a 100 nm thermal SiO_2 layer grown on a 200mm silicon substrate. A 70nm TiN layer was sputtered form a Ti target in a N_2 atmosphere; following this a 1-2nm nominal Ni or Co film was deposited by PVD (Endura, Applied materials). For CNT via growth, contact hole arrays (150-300nm) were etched in SiO_2 (300nm) on a TiN (70nm) bottom electrode common to all via holes to mimic the blanket substrates. Details of the via fabrication process can be found elsewhere [10]. CNT growth was carried in a 200mm microwave (2.45 GHz) PECVD chamber (TEL, Japan). The microwave plasma was located remotely from the wafer surface to avoid excessive ion-bombardment. Charged species could be further filtered by means of an electrically biased ion reflector located between the wafer and plasma source. In a typical CNT growth experiment the catalyst film was transformed into active metal nanoparticles in a H_2 or NH_3 plasma for periods of up to 5 min. C_2H_4 was charged into the chamber and CNTs were grown for periods of up to 30mins. To further control the active carbon species during growth a bias (-/+100V) was applied to the ion reflector. The as-grown CNTs were analyzed using SEM (Hitachi, SU8000 FE), TEM (Tecnai F30 ST (FEI) with a 300 kV FEG). CNT bundles grown within vias were removed using a tungsten nanoprobes (Kleindiek nanomanipulators).

DISCUSSION

Figure 1 outlines the particle distributions determined from SEM formed from a 1 Ni thin film on TiN under differing plasma annealing conditions and gas ambient (NH_3 and H_2). An upper temperature limit of 540 °C is used as a reference since dense aligned CNT growth was routinely obtained see figure 3(a) and (b).

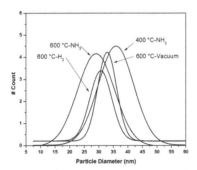

Figure 1:   Ni nanoparticle size distributions formed by annealing a 1nm Ni film on TiN at 400 °C and 540 °C in different gas ambient

At 400°C the diameter distribution ranges from 20nm to 50nm centered at 37nm approx. At the upper end of this scale it can be seen that the film has not de-wetted completely into particles and large island like droplets are formed. By increasing the temperature (540 °C) we observe a smaller average particle size ($NH_3 = 25nm$) and a narrower distribution. In fact, it is known that if the surface energies (J/m^2) of the substrate and the catalyst are similar ($TiN;1.6$, Ni; 2.08) [13-14] it is less easy to de-wet and stabilize the catalyst by annealing alone as compared with lower surface energy oxides ($Al_2O_3 \approx 0.1$) [15-16].

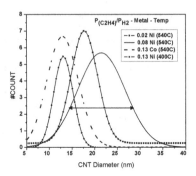

Figure 2:   CNT diameter (nm) for Ni/TiN grown under different C_2H_4/H_2 ratios at 540°C and 400°C included also is a CNT distribution from a Co/TiN starting layer

The average CNT diameter has reduced to 18nm for Ni/TiN from a particle size of 27nm (540°C) and for Co/TiN shows a reduced average size of 11.5nm. By reducing the temperature to 400°C a significant reduction to 12nm for Ni/TiN is observed. Indicating that at 400°C with an appropriate carbon concentration Ni is an effective CNT catalyst. A SEM view of the Co/TiN sample is shown in figure 3 (a) with an average length 4.5 um this image is taken from a completely uniform CNT layer covering a 200mm wafer surface at 540 °C. A via sample with selective CNT growth is depicted in figure 3 (b) here CNTs are grown selectively form a localized Ni/TiN layer at the bottom of the via with densities in the order of $\sim 10^{11}$ CNTs/cm^2, up to lengths of 4 um, and an average diameter of 8nm. Although, above the CMOS thermal budget to our knowledge this is comparable to state of the art CNT via growth on TiN [17]. The reduced diameter in this instance can be attributed to a) less catalyst material reaching the via bottom during the PVD sputtering, thus smaller particle size and b) side wall growth as catalyst may adhere to the sidewall during sputtering [18]. The selective CNT growth in vias highlights the catalyst to substrate interaction in that the top surface of the via sample or field is a SiC layer surface - used as a stop layer for CMP further in the integration and testing process [19] - and no CNT growth is observed. In fact under the annealing and CNT growth conditions developed for (Ni or Co)/TiN we observe large island like Ni formation on the SiC top surface forming a graphitic like layer that can easily be removed post growth. This selectivity can be advantageous in that a dense CNT layer on the top surface may hinder gas species reaching the bottom of the via, possibly reducing CNT yield and density. Returning to particle size vs CNT size reduction, the change in particle size is most likely attributed to further metal reduction during CNT growth, temporary immobilization on TiN by encapsulation form the initial carbon layers on the nanoparticle surface [19] and eventual lift off during tip growth. Indeed for low temperature CNT growth with Ni metal can be seen not only at the tip but extruded through the lumen of the CNT (data not shown). Furthermore, in many cases on TiN vertical alignment is not observed possibly due to large catalyst size distributions. The current data suggest for a particular size distribution with a difference greater than approximately 15nm (see arrow in figure 2) CNTs grow at different rates (enhanced at 540°C) and spaghetti-like or unaligned layers are grown. Indeed the two curves to the left in figure2 are representative of aligned CNTs (figure 3(a) and (b)) while the two curves to the left are of unaligned layers (SEM data not shown).

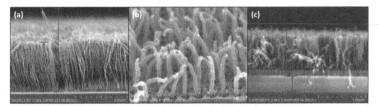

Figure 3   Dense vertically aligned CNT growth from (a) a Co(1nm)TiN layer fully uniform over a 200mm Silicon wafer surface (b) Selective CNT growth from a Ni(1nm)TiN layer at the bottom of a via patterned 200mm wafer and (c) on Ni/TiN 400°C

Figure 4 shows CNT bundles grown within a via hole array at 400°C using a biased ion reflector positioned between the gas stream and wafer surface. A 100% yield was maintained throughout the via hole arrays (300 – 160nm). The ion filter can control the amount of carbon radicals to ionizing species in gaseous stream. In this instance it appears that a high concentration of carbon species has been brought in contact with the metal catalyst layer at the via bottom. The low temperature (400 °C) limits fast CNT growth as mentioned in relation to C-solubility in bulk metals [5] and a graphitic cap with a diameter equal to the via is preferentially formed as shown in fig 4 (b). Interestingly, no deviation in CNT length with via diameter was observed and the CNT height above the via was controlled to 600 $\pm$ 30 nm. From the level of via filling observed we can estimate (SEM) the density of the CNT bundle to be in excess of 10^{11} CNTs/cm^2.

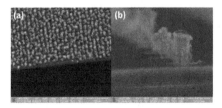

Figure 4:   CNT bundles form Ni TiN grown at 400 °C using a ion reflector biased to - 100V(a) tilted SEM indicating high level of via filling and (b) cross-sectional image showing one individual CNT bundle

A closer examination of the CNT bundle was carried out by removing the bundles using nanoprobe manipulation as outlined in the image sequence in figure 5 (a-d). To our knowledge this strange morphology has not been observed in CNT via reports and resembles a carbon nanostructured 'jelly-fish'. The location of the Ni catalyst can be identified from the EDS map in (b). The carbon (green) cap is pointing to the left of the CNT bundle attached to the W probe. Carbon data (green) on single CNTs show that drift was well controlled in these measurements the carbon and nickel (red) on the tungsten tip results from brehmsstrahlung background and does not indicate existence of these elements on the tip. Nickel seems to be concentrated in particles (or clusters of particles) in the bundle. The sides of the cap also have high concentrations of nickel. This is possibly caused by particles from the via sidewalls trapped in the cap when bundle growth pushes the cap upwards. The nanostructure of an individual CNT is shown in (d) with an ordered graphitic wall structure as a high level of graphitization is required to maximize the number of conducting channels within an individual via. Ideally for CNT interconnects an effective contact should be established to every CNT shell (wall) within a via. In this case the graphitic gap may enhance this electrical contact. Nevertheless, as the bundle protrudes over the via, the cap will be removed in a subsequent CMP planarization step before the formation of a top metal contact layer.

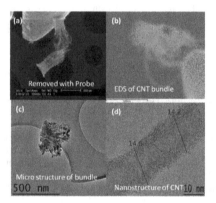

Figure 5: (a) SEM image of CNT buFndles from figure 4 attached to a W nanoprobe tip, (b) overlayed EDS data carbon (green) and Nickel (red) indicating catalyst position transferred to a TEM grid (c) and HRTEM image indicating nanostructure of a typical MWCNT

CONCLUSIONS

We have demonstrated the growth of CNTs on a metallic –like underlayer; TiN, at 400 °C mindful of the constraints imposed by CMOS processing. The growth process could be directly transferred to via hole arrays where selective CNT growth was achieved. To realize CNT growth at 400°C the use of a remote microwave plasma producing active C growth species was required. Ion filtering of the carbon plasma could completely fill vias with a mixture of CNTs and a graphitic-like cap. From a experimental viewpoint the challenges facing CNTs as interconnects have been outlined: namely, a) maintaining a uniform small diameter catalyst distribution on a metal-like under layer, this controls alignment and hence density, b) maintaining a high level of graphitization in the individual CNT structure at 400 °C.

Acknowledgements
This work was partly supported by the EU project CARBonCHIP (NMP4-CT-2006-016475) and METACEL project (IWT-SBO project 060031), funded by IWT-Vlaanderen.

REFERENCES

1. A. R. Harutyunyan, G. Chen, T. M. Paronyan, E. M. Pigos, O. A. Kuznetsov, K.Hewaparakrama, S. M. Kim, D. Zakharov, E. A. Stach, and G. U. Sumanasekera, Science, **326**, 116, (2009)

2. A. Romo-Negreira, D.J. Cott, A. S. Verhulst, S. Esconjauregui, N. Chiodarelli, J.Ek-Weis, C. M. Whelan, G. Groeseneken, M. M. Heyns, S. De Gendt, P. M. Vereecken., Mater. Res. Soc. Symp. Proc., **1079E**, N06-01, 2008.

3. Akio Kawabata, Shintaro Sato, Tatsuhiro Nozue, Takashi Hyakushima, Masaaki Norimatsu, Miho Mishima, Tomo Murakami, Daiyu Kondo, Koji Asano, Mari Ohfuti, Hiroshi Kawarada, Tadashi Sakai, Mizuhisa Niheiand Yuji Awano, IEEE 987-1-4244-1911-1/08 (2008)

4. S. Armini and P. M. Vereecken, Abstract No 2780, ECS Meeting Abstracts, **902**, Vienna, Austria, Oct 4–9, (2009)

5. R. T. K. Baker, P. S. Harris, R. B. Thomas and R. J. Waite, J. Catal. **30**, 86 (1973)

6. V. I. Merkulov, D. H. Lowndes, Y. Y. Wei, G. Eres and E. Voelkl Appl. Phys. Lett., **76**, 24, 12 (2000)

7. Guangyu Zhang, David Mann, Li Zhang, Ali Javey, Yiming Li, Erhan Yenilmez, Qian Wang, James P. McVittie, Yoshio Nishi, James Gibbons, and Hongjie Dai, PNAS, **102**, 45, 16141–16145, (2005)

8. Cecilia Mattevi, Christoph Tobias Wirth, Stephan Hofmann, Raoul Blume, Mirco Cantoro, Caterina Ducati, Cinzia Cepek, Axel Knop-Gericke, Stuart Milne, Carla Castellarin Cudia, Sheema Dolafi, Andrea Goldoni, Robert Schloegl and John Robertson, J Phys. Chem. C   112 , **32**,12207-12213 (2008)

9. Guofang Zhong, Takayuki Iwasaki, John Robertson, and Hiroshi Kawarada, J Phys Chem B letters, **111**, 1907-1910, 2007

10. Nicoló Chiodarelli, Kristof Kellens, Daire J. Cott, Nick Peys, Kai Arstila, Marc Heyns, Stefan De Gendt, Guido Groeseneken, and Philippe M. Vereecken, J Electrochem. Soc, **157**, 10, K211-K217, (2010)

11. Tatsuki Hiraoka, Takeo Yamada, Kenji Hata, Don N. Futaba, Hiroyuki Kurachi, Sashirou Uemura, Motoo Yumura and Sumio Iijima, J. Am. Chem. Soc.,**128** (41), 13338–13339, 2006

12. Teresa de los Arcos, M. Gunnar Garnier, Peter Oelhafen, Daniel Mathys, Jin Won Seo, Concepcion Domingo, Jose Vicente Garcia-Ramos, Santiago Sanchez-Cortes, Carbon **42** (2004) 187–190

13. M Marlo, V Milman, Phys Rev B **62** 2899 2000

14. W. R. Tyson and W. A. Miller, Surf. Sci. 62, **267**, 1977

15. Z. Łodziana, N.-Y. Topsøe, and J. K. Nøskov, Nature Mater. 3, **289**, 2004

16. C. Zhang, Yan, C. S. Allen, B. C. Bayer, S. Hofmann,B. J. Hickey, D. Cott, G. Zhong and J. Robertson, JOURNAL OF APPLIED PHYSICS **108**, 024311 (2010)

17. Daisuke YOKOYAMA, Takayuki IWASAKI, Kentaro ISHIMARU, Shintaro SATO, Takashi HYAKUSHIMA, Mizuhisa NIHEI, Yuji AWANO, and Hiroshi KAWARADA Jpn. J. Appl. Phys., **47**, 4 (2008)

18. Xiaoxing Ke, Sara Bals, Daire Cott, Thomas Hantschel, Hugo Bender, and Gustaaf Van Tendeloo Microsc. Microanal. **16**, 210–217, 2010

19. N. Chiodarelli, Y. Li, D. J. Cott, S. Mertens, N. Peys, M. Heyns, S. De Gendt, G. Groeseneken, P. M. Vereecken, Integration and electrical characterization of carbon nanotube via interconnects, Microelectronic Engineering (2010) - in publication

20. Yuichi Yamazaki, Masayuki Katagiri, Naoshi Sakuma, Mariko Suzuki, Shintaro Sato, Mizuhisa Nihei, Makoto Wada, Noriaki Matsunaga, Tadashi Sakai, and Yuji Awano, Applied Physics Express **3** (2010) 055002

**CNTs Growth, Exploring Novel CNT Growth Techniques
and Growth Mechanisms II**

Mater. Res. Soc. Symp. Proc. Vol. 1284 © 2011 Materials Research Society
DOI: 10.1557/opl.2011.220

On the Formation of Carbon Nanotube Serpentines: Insights from Multi-Million Atom Molecular Dynamics Simulation

Leonardo D. Machado[1], Sergio B. Legoas[2], Jaqueline S. Soares[3], Nitzan Shadmi[4] , Ado Jorio[3], Ernesto Joselevich[4], and Douglas S. Galvao[1]
[1]Applied Physics Department, State University of Campinas, Campinas-SP, 13083-459, Brazil.
[2]Physics Department, Federal University of Roraima, Boa Vista-RR, 69304-000, Brazil.
[3]Physics Department, Federal University of Minas Gerais, Belo Horizonte-MG, *30123-970*, Brazil.
[4]Department of Materials and Interfaces, Weizmann Institute of Science, Rehovot, 76100, Israel.

ABSTRACT

In this work we present preliminary results from molecular dynamics simulations for carbon nanotubes serpentine dynamics formation. These S-like nanostructures consist of a series of parallel and straight nanotube segments connected by alternating U-turn shaped curves. Nanotube serpentines were experimentally synthesized and reported in recent years, but up to now no atomistic simulations have been carried out to address the dynamics of formation of these structures. We have carried out fully atomistic molecular dynamics simulations in the framework of classical mechanics with a standard molecular force field. Multi-million atoms structures formed by stepped substrates with a carbon nanotube (about 1 micron in length) placed on top of them have been considered in our simulations. A force is applied to the upper part of the tube during a short period of time and then turned off and the system set free to evolve in time. Our results showed that these conditions are sufficient to form robust serpentines and validate the general features of the 'falling spaghetti mechanism' previously proposed to explain their formation.

INTRODUCTION

Carbon nanotube serpentines (CNSs) are S-like nanostructures composed of regularly spaced and parallel straight segments, connected by alternating U-turn shaped curves. These remarkable structures have been experimentally obtained growing long carbon nanotubes on sapphire and quartz patterned substrates under the presence of a flow gas flux [1,2,3,4]. CNSs were firstly synthesized in 2008 by the Joselevich's group [1]. Recently, other groups have reported similar results [2-5].

CNS formation has been qualitatively explained based on the "falling spaghetti mechanism" [1]. The serpentines would be formed in a two-step process, where the isolated nanotubes are first grown standing up from the stepped substrates, and at a second stage, the tube would fall down preferentially along the steps, creating the oscillatory patterns, like spaghetti falling on a tilted bamboo mat [1]. The force that would be primarily responsible for the tube fall is the strong nanotube-surface interactions (mainly van der Waals forces). In this case, the growing nanotube is buoyant over the substrate and is submitted to a tension by the action of a gas flow perpendicular to the steps. After starting to fall the nanotube initiates an oscillatory motion with the tube being adsorbed by the substrate in a sequence of straight segments along the steps or lattice direction, connected by U-turn curves. CNS formation would thus be the result of a competition between the flow-induced tension and the elastic deformations due to strong tube-surface adhesion [1].

In all the serpentine experimental realizations some common features are present. One is a substrate with appropriate topological surface format (in most cases, step patterned topologies). Another necessary condition is a forward gas flow action in order to produce a translational movement to the falling/landing tube. In some realizations thermal gradients play an important role modulating the buoyancy force and controlling the nanotube landing [4]. If the carbon tube is long enough, serpentines will spontaneously be formed as a result of the appropriate balance of these aspects.

In the present work, we report preliminary results from molecular dynamics simulations of the dynamical processes of the transformation of long single wall carbon nanotubes (placed on stepped substrate and submitted to an external force) into serpentines.

MODELING

We have carried out fully atomistic molecular dynamics simulations, in the framework of classical mechanics with standard molecular force field [5], using the parallel molecular dynamics NAMD code [6] on its CUDA implementation [7]. Quartz and graphite stepped substrates were considered (Fig. 1a). Nanotubes (about 1 micron in length) are then placed on top of these substrates, with and without the presence of a nanoparticle on its free end (Fig. 1b). A forward force was applied to the upper part of the tube along the x direction (Fig. 1a) during a short period of time, then turned off and the system set free to evolve in time. If the applied force would be turned on all the time this would require longer tubes and larger substrates because of the acquired kinetic energy.

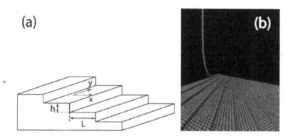

Figure 1. Schematic illustration of a step patterned substrate considered in our study. The forward force is applied parallel to the x direction and perpendicular to y ones. h and L are the step dimensions. Values ranging from 4 up to 5 Å for h, and from 30 up to 32 Å for L, were used in our calculations. (b) Snapshot of a used quartz stepped substrate, with a (6,0) CNT standing on top of it.

In all the simulations the following procedures were used. Initially, the substrates are modeled from large SiO_2 or graphite slabs to produce the stepped structures (Fig. 1a). We have used structural models containing about 1.5-2.0 million atoms. Next, a long single walled carbon nanotube (about 1.0 micron length) is built. We have considered (6,0) and (12,0) nanotubes, corresponding diameters of 4.7 and 9.4 Å, respectively. The tubes are bent into a L-shape with straight sections of ~ 350 and ~ 10000 Å. The L-tube is placed on the top of the substrates in a way that its shorter section is along a step (Fig. 1b). The system is then geometrically optimized.

In order to mimic the massive (catalytic) particle present during CNT growth, the mass of the free CNT upper end was attributed to be ~5 500 a.m.u. In all the calculations, the substrate had its atomic positions fixed. In all the molecular dynamics calculations we have considered the initial temperature to be at 300 K in the microcanonical ensemble. An external forward force (perpendicular to the step substrate direction, see Fig. 1a) was applied to the suspended tube (0.7 pN/atom) during 36 ps, then turned off and the tube dynamics recorded. After some exploratory simulations these range values were found to be effective to produce well-formed serpentines.

RESULTS AND DISCUSSION

In Figs. 2a-e we show typical snapshots from molecular dynamics simulations where the serpentine formation was observed. The indicated time (in pico seconds) is with relation to the elapsed time since the application of the external force. We have observed that, without the use of the external force, the long tube falls on the substrate and on itself, in an irregular form, and no serpentine is formed.

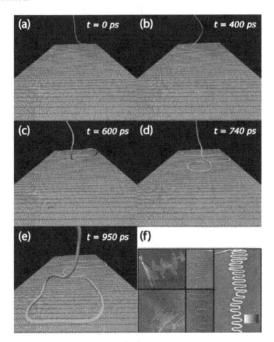

Figure 2. Panels (a)-(e) show typical snapshots from molecular dynamics simulations of carbon nanotube serpentine formation on a stepped quartz substrate. (f) Some typical SEM experimental images of serpentines grown on quartz substrates at different flow rates [1]. Typical dimensions of the substrates are 1000 Å x 2000 Å.

In contrast, when the forward force is applied to the falling tube (in our case, during 36 ps), the nanotube falls along the steps and exhibit rapid oscillatory motions. While the part of the tube that is in contact with the surface will lead to well and spontaneously formed serpentine-like structures (Figs. 2c-d), the remaining suspended part of the tube continues to exhibit random motion (Fig. 2e). These results seem to be consistent with the experiments. The obtained serpentine structures (Fig. 2e) reproduce quite well the general features observed in the experiments (Fig. 2f).

For the same type of tube, the separation between the U-turn segments can, in principle, be controlled by varying the applied force (larger force values would generate larger segment separations). The obtained non-uniformity of these segments (again consistent with what is observed in the experiments, Fig. 2f) is consequence of kinetic/thermal fluctuations at nanoscale and much more difficult to control. In the experiments the non-uniformity can be attributed, among other factors, to the fluctuations in the gas flux (intensity and directions) [1-4]. We have also investigated whether the nanoparticle at the end tube has an active role on these fluctuations. We run simulations with and without the presence of these particles. Our results showed that the particle plays indeed an active role in the serpentine formation. It helps to damp large amplitude tube oscillations that significantly contribute to prevent the formation of more uniformly shaped serpentines.

In order to investigate how important is the substrate to serpentine formation, we have also considered the same type of simulations but using stepped graphite substrates instead of quartz ones. Our results showed that the formation dynamics is slightly different, well-formed serpentines are also possible on graphite substrates [9] for the tubes investigated here, ((6,0) and (12,0)). It seems that the steps are much more important to induce serpentine formation than the materials that compose the substrates. Also, the serpentine formation was not observed to be very sensitive to h and L (Fig. 1a) values.

In order to better understand the nanotube structural changes during the process of serpentine formation we calculated the temporal evolution of the tube internal strain forces. In Fig. 3 we show typical results for the case of a (6,0) carbon nanotube.

Topologically, the tube can be considered as formed by a series of six-atom rings interconnected by sp^2 bonds. We have calculated the bond interactions between two neighboring rings, as for instance, ring$_i$ and ring$_j$ (Fig. 3a). In the rings some atoms are submitted to tension forces (positive value F_{11}), while others to compression ones (negative value F_{44}) (Fig. 3a). The forces acting on the atoms have an oscillatory behavior, due mainly to elastic deformations (as a consequence of interactions with the substrate) and thermal fluctuations. At the moments just before the formation of a given U-turn, the forces on certain atoms can substantially increase. In Fig. 3b shows typical results for time evolution of the stress for the atoms of a given ring. this case, we present the results for two atoms (labeled as 1-1 and 4-4 in panel (a) of Fig. 3). We can see the fast variation of the force acting on a single atom. In order to obtain more precise information about the force profiles, we used a fast Fourier transform to smooth the data (Fig. 3b). As we can see from the Figs. some atoms can experience stress forces greater than 30 kcal/mol/ Å (> 2 nN).

A more comprehensive view of the force profiles of the first U-turn formation on a graphite substrate (the critical process for serpentine formation) is shown in Fig. 4, where results as the ones indicated in Fig. 3 are presented for three different tube regions at different times. At the initial time, the external forward force is applied to the tube, and 360 ps later the first U- turn is formed. In (a) we can see the force variation for the ring labeled by *a* in panels (1) (lateral view) and (2) (front view).

The force on each of the six atoms of the analyzed rings varies between ~ ±10 kcal/mol/Å. However, between ~ 90 and ~ 190 ps, the forces increase on all the atoms up to ~ 35 kcal/mol/Å, at the moments of U-turn formation. This phenomenon is observed in the other rings at different times. As the tube falls on the substrate, an elastic deformation wave is created and it propagates along the tube (see panel (b) and (c) in Fig. 4). The stress values of the tube segments parallel to the steps and in contact with it return to their previous values after the U-turn formation. The process is repeated at each U-turn formation, leading to the serpentine formation.

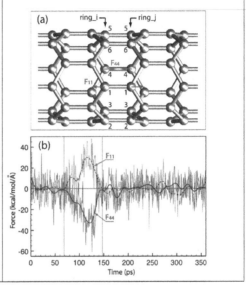

Figure 3. (a) Schematic view of a section of the (6,0) nanotube. The stress force along the tube is calculated by the bond interaction between pairs of atoms labeled 1-1, 2-2, etc. In (b) it is presented typical results for the time evolution of the internal stress of the nanotube for a particular ring. Vertical lines (in red color) indicate the interval of time when the ring atoms are experiencing maximum stresses.

From the simulations and force profile analysis it is possible to explain how the serpentines are formed. The process involves a balance of different kind of forces, elastic deformations, stress-strain force distributions modulated by the materials and format of the substrate steps. As the forward force if applied the tube starts to move forward, but at the same time the interactions with the substrate (mainly van der Waals forces) pulls it down toward the substrate. As the tube segments starts to interact with the substrate elastic waves (deformations) are generated and propagate through the tube which tends to align it with the substrate steps. This continues until the elastic limit (maximum stress) is reached (which depends on multiple factors, such as: kind of substrate, temperature, applied external force, catalytic particle, etc.) and the forward tube force/velocity overcomes the elastic deformation, leading to a U-turn formation. The repetition of these processes leads to serpentine formation. From the simulations we observed that, as far as the top part of the tube continues to be ahead of its main body, serpentine-like structures can be formed. When this condition is not satisfied it tube falls on itself, producing looped or ill-formed serpentines. Interestingly, the simulations showed that, although complex and involving many factors, the qualitative general trends of the serpentine formation are basically the ones of the proposed 'falling spaghetti mechanisms' [1]. Further studies to determine the importance of other factors that were not investigated in this work are in progress.

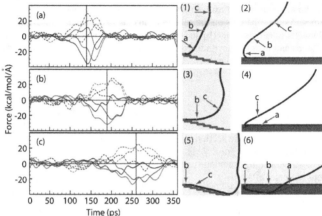

Figure 4. Temporal evolution of the nanotube stress profile during the serpentine formation on a graphite substrate. In (a), (b) and (c) are shown the forces on three different points of the nanotube. In panels 1-6 we present snapshots of the corresponding tube configurations (lateral and front views) on its first U-turn formation at the times indicated by the vertical lines in a-c. Labels a, b and c in the panels 1-6 correspond to the ring positions where the forces were calculated. Same color pair (full and dashed) curves refer to specific rings, as shown in Fig. 3).

ACKNOWLEDGMENTS

This work was supported by the Brazilian agencies CNPq, FINEP, FAPEMIG and FAPESP. EJ and NS acknowledge support from E.J. acknowledges support from the Israel Science Foundation, the US-Israel Binational Science Foundation, the Kimmel Center for Nanoscale Science, and the Legrain, Djanogly, Alhadeff, and Perlman Family foundations.

REFERENCES

1. N. Geblinger, A. Ismach, and E. Joselevich, *Nature Nanotechnology.* **3**, 195 (2008).
2. S. Jeon, C. Lee, J. Tang, J. Hone and C. Nuckoll, *Nano Res* **1**, 427 (2008).
3. J. Huang and W B. Choi, *Nanotechnology* **19**, 505601 (2008).
4. Y. Yao *et al. Adv. Mater.* **21**, 4158 (2009).
5. J. S. Soares *et al. Nano Lett., ASAP doi:10.1021/nl103245q.*
6. A. D. MacKerell *et al. J. Phys. Chem.* B **102**, 3586 (1998).
7. J. C. Phillips *et al. J. Comput. Chem.* **26**, 1781 (2005). NAMD, http://www.ks.uiuc.edu/Research/namd/.
8. J. E. Stone *et al. J. Comput. Chem.* **28**, 2618 (2007).
9. L. D. Machado, S. B. Legoas, J. S. Soares, N. Shadmi, A. Jorio, E. Joselevich, and D. S. Galvao, to be published.

Electronic, Optical, and Magnetic Properties of Carbon Nanomaterials I

Mater. Res. Soc. Symp. Proc. Vol. 1284 © 2011 Materials Research Society
DOI: 10.1557/opl.2011.646

Graphene for Magnetoresistive Junctions

J. Inoue[1], T. Hiraiwa[1], R. Sato[1], A. Yamamura[1], S. Honda[2], and H. Itoh[3]

[1]Department of Applied Physics, Nagoya University, Nagoya 464-8603, Japan

[2]ORDIST, Kansai University, Suita 564-8680, Japan

[3]Department of Pure and Applied Physics, Kansai University, Suita 564-8680, Japan

ABSTRACT

Influence of the linear energy-momentum relationship in graphene on conductance and magnetoresistance (MR) in ferromagnetic metal (FM)/graphene/FM lateral junctions is studied in a numerical simulation formulated using the Kubo formula and recursive Green's function method in a tight-binding model. It is shown that the contribution of electron tunneling through graphene should be considered in the electronic transport in metal/graphene/metal junctions, and that the Dirac point (DP) is effectively shifted by the band mixing between graphene and metal electrodes. It is shown that MR appears due to spin-dependent shift of DP or spin-dependent change in the electronic states at DPs. It is shown that the MR ratio caused by the latter mechanism can be very high when certain transition metal alloys are used for electrodes. These results do not essentially depend on the shape of the junction structure. However, to obtain high MR ratios, the effects of roughness should be small.

INTRODUCTION

Magnetoresistance (MR) is a key phenomenon in the field of spintronics. The giant MR (GMR) [1,2] in ferromagnetic metal (FM)/non-magnetic metal/FM junctions and the tunnel MR (TMR) [3,4] in FM/insulator/FM junctions have been widely used for technological applications such as in magnetic sensors and magnetoresistive memories. In GMR, spin-dependent scattering at the interfaces of junctions plays an essential role [5], whereas in TMR the crucial phenomena are spin polarization in FMs and spin-dependent decay rate in insulating spacers [6,7]. Currently, the search for novel combinations of FM and non-magnetic spacers is required to develop advanced spintronics devices. Among various spacer materials, graphene is one of the most attractive for device applications because of its high mobility, long spin-diffusion length, and a planar lattice structure [8-10].

Graphene, which is composed of a two-dimensional honeycomb lattice of carbon atoms, has a characteristic electronic structure near the Fermi level: its conduction and valence bands meet at only two states called Dirac points (DPs) in the Brillouin zone, and its energy momentum dispersion is linear near the DPs. In other words, graphene is a gapless semiconductor in which electrons behave as massless electrons. The electronic structure of graphene is somewhat different from that of conventional non-magnetic metals, semiconductors, and insulators, and therefore, may produce novel features of conductance in junctions when used as a spacer material. It is also expected that the graphene spacers may lead to a novel mechanism of MR in FM/graphene/FM lateral junctions, distinct from those in GMR and TMR junctions. So far, a few experiments have been conducted to measure the MR ratio for FM/graphene/FM lateral junctions [11-13]. The observed MR ratios, however, are not sufficiently large for device applications, and a clear mechanism for the MR is lacking. Therefore, theoretical study of MR in two-terminal graphene junctions with FM electrodes is desirable to clarify the mechanism and provide a guiding principle for junction design.

The objective of this work is to clarify the role of linear energy dispersion on conductance in FM/graphene/FM junctions and to discuss its effects on MR in FM/graphene/FM junctions. For this purpose, we perform numerical calculations of conductance in FM/graphene/FM lateral junctions having zigzag edge contacts. Because K' point, one of the two DPs, is included in the momentum states parallel to the junction contact in this geometry, we can analyze the property of the electrical conductance of junctions in detail. A portion of the numerical results of MR has been published in our previous reports [14, 15].

THEORY

Schematic figures of FM/graphene/FM lateral junctions with zigzag edge contacts are shown in Figure 1. Two models are adopted for the electrodes: a bcc Fe lattice (figure 1(a)) with three atomic layers, and an fcc Ni lattice with four atomic layers (figure 1(b)). The electronic structures of graphene and the electrodes are described by the sp^3 orbitals and full orbital TB model, respectively, although the p_z orbital is essential for the graphene layer. A periodic boundary condition is used along the y direction, and the graphene length L (along the x direction) is less than ~5000 monolayers (MLs), where the number is considered in such a way that one zigzag chain corresponds to two MLs. Since a periodic boundary condition is adopted along y direction, the momentum $k_\parallel$ parallel to the contact is conserved. The band parameters of the TB model used by us are taken from textbooks [16, 17]. The conductance of the junction is calculated using the Kubo formula with recursive Green's function method. Details are described in Itoh and Inoue [18].

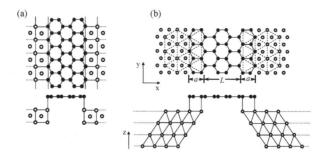

Figure 1. Structures used for (a) bcc Fe/graphene/bcc Fe junctions, and (b) fcc Ni/graphene/fcc Ni junctions. Closed circles are C atoms, and open circles are bcc Fe or fcc Ni atoms. L and a indicate length (ML) in graphene layer and overlapping region between graphene and electrodes, respectively. $L = 6$ and $a = 3$ MLs in this figure.

RESULTS and DISCUSSION

We have reported that the following characteristic features can be seen in the momentum ($k_{\parallel}$) resolved conductance $\Gamma(k_{\parallel})$ in graphene junctions [8]: (1) The state with maximum conductance e^2/h, which may be identified as the DP, shifts with graphene length L and the hopping integrals between graphene and electrodes. The state at which the conductance attains maximum value may be called an effective DP, and the momentum at this point is indicated as k_{DP} hereafter. (2) Conductance near the DP is not zero but decays as $\exp(-c|\,k_{DP} - k_{\parallel}|L)$ with a constant c.

The exponential dependence of $\Gamma(k_{\parallel})$ is interpreted as follows. Because the energy band gap of graphene closes at only two DPs in the Brillouin zone, graphene has two conductive channels at the Fermi energy but the other states are insulating. However, since energy dispersion is linear near the DPs, the energy band gap can be infinitesimally small as $k_{\parallel} \to K'$. Therefore, tunnel conductance does not decay exponentially with increasing barrier thickness (graphene length in this case) contrary to that in typical tunnel junctions, but shows different thickness dependence given by $\Gamma(L) \propto 1/L$. We have attributed the shift in the effective DP to a change in the electronic states near the contact of metal/graphene/metal junctions by calculating the energy bands of finite-size systems in a tight-binding model (p_z orbital in graphene). Details will be published elsewhere.

These two features can be seen in figure 2(a), which shows momentum resolved conductance $\Gamma(k_{\parallel})$ calculated for bcc Fe/graphene/bcc Fe junctions. Broken and dotted curves show $\Gamma(k_{\parallel})$ of majority and minority spin states in parallel (P) alignment of bcc Fe magnetizations, respectively, and a solid curve indicates $\Gamma(k_{\parallel})$ in anti-parallel (AP) alignment of bcc Fe magnetizations. The momentum k_{DP} with maximum conductance e^2/h shifts from $\bar{K}'$ point shown in the figure, and $\Gamma(k_{\parallel})$ decays exponentially with increasing $k_{DP} - k_{\parallel}$. Here, $\bar{K}'$ indicates the momentum of K' state along $k_{\parallel}$. Spin-dependent shift of k_{DP} occurs because the band mixing between graphene and the spin-polarized Fe band in the electrodes is spin-dependent.

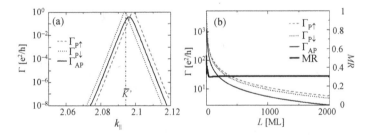

Figure 2. (a) Calculated results of momentum resolved conductance Γ for bcc Fe/graphene/bcc Fe junctions with graphene length $L = 1000$ MLs. Broken and dotted curves are results of Γ for majority ($\uparrow$) and minority ($\downarrow$) spin states in parallel alignment of bcc Fe magnetizations, and solid curve shows those in anti-parallel alignment. (b) Calculated results of Γ and MR ratio as a function of L.

The spin-dependent shift of the effective DP caused by band mixing between graphene and the electrodes plays an important role in MR. The maximum value of $\Gamma(k_\parallel)$ in AP alignment is smaller than that in P alignment as shown in figure 2(a), because the shift of the effective DP is in conflict with the majority and minority spin states in the AP magnetization alignment. As a result, the MR appears. Figure 2(b) shows calculated results of conductance Γ and MR as a function of graphene length L. Γ decays as $1/L$ as mentioned, and MR appears almost independently of L. Here, the MR ratio is defined as $MR = (\Gamma_P - \Gamma_{AP})/(\Gamma_P + \Gamma_{AP})$. Because the shift in the effective DP depends on the magnitude of band mixing between graphene and electrodes, the contact should be metallic to obtain a sufficient spin dependence of the effective DP shift. For junctions with a tunnel barrier between graphene and an electrode, there may not be any shift in the DP, resulting in an absence of MR.

We have pointed out that other mechanisms of MR are possible in FM/graphene/FM junctions [8], namely, a matching/mismatching mechanism of the DP with conduction states in the electrodes, and a spin-dependent change in the electronic structure near DP. The former mechanism is the simplest for MR in FM/graphene/FM junctions. Since the DP is the only available conduction channel, spin-dependent matching/mismatching of the conductive states in graphene and electrodes leads to a high MR ratio. The MR effect appears only when the $\bar{K}'$ point is included in either the majority or minority spin conduction state of the electrode, and we expect that the MR ratio is either 1 or 0. However, the mechanism is very simple for junctions with realistic ferromagnetic metals. Nevertheless, this kind of mechanism is possibly important when the electronic structure near the DP is strongly spin dependent. When the bcc Fe bands are shifted by a certain amount, local spectral density of states (DOS) near $\bar{K}'$ becomes very small at the contact of graphene and fcc Fe layers [15].

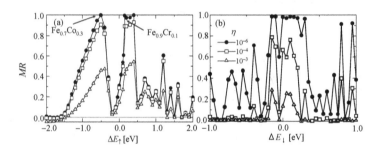

Figure 3. Calculated results of MR ratio for (a) bcc Fe/graphene/bcc Fe and (b) fcc Ni /graphene/fcc Ni junctions obtained by shifting the energy bands of majority spin state in bcc Fe or shifting the minority spin bands in fcc Ni. Corresponding alloy compositions are shown in the figure. Graphene length is 1000 MLs for both junctions. η indicates life-time broadening included in Green's function used in the formalism. Overlapping region a in fcc Ni/graphene/fcc Ni junctions are 50 MLs for both contacts.

The negligible local spectral DOS near the effective DP may result in a high MR ratio because the matching of the conductive states in graphene and the electrodes becomes worse. Figure 3(a) shows calculated results of MR ratios as a function of a shift of the majority (↑) spin band. Corresponding composition of Fe alloys is presented in the figure. We find that a huge MR effect may appear in junctions with electrodes made of FeCo, FeCr, and FeV alloys. The result is attributed to a suppression of conducting states in the majority spin state near the junction contacts, which indicates that matching between DPs and conduction channel of electrodes becomes worse in the spin state. We also present the results of MR ratios when a life-time broadening of DOS is taken into consideration. With increasing life-time broadening, the MR ratio decreases since it broadens the spectral DOS in general. Because the life-time broadening corresponds to electron scattering by roughness, we must conclude that the MR ratio decreases with increasing roughness in the graphene layer.

Figure 3(b) shows similar results for fcc Ni/graphene/fcc Ni junctions, for which the junction structure shown in figure 1(b) is utilized. The overlapping region a at the left and right contacts are both 50 MLs. Contrary to bcc Fe/graphene/bcc Fe junctions, MR ratio is somewhat stable against the band shift of down (↓) spin state of fcc Ni. This means that Ni rich Ni-Co or Ni-Cu alloy electrodes should show high MR ratio. The results are insensitive to the area of the overlapping region. Thus, we may conclude that the lattice structure of the electrodes and the shape of contacts have essentially no impact on the features of conductance in FM/graphene/FM junctions. Quantitatively, however, fcc Ni/graphene/fcc Ni junctions may be promising to obtain high MR ratios.

CONCLUSIONS

The linear energy-momentum relationship near the Fermi level in graphene strongly influences the conductance and MR in FM/graphene/FM lateral junctions. Two important findings are reported. First, the contribution of electrical tunneling should be considered properly in the calculation of the electronic transport in metal/graphene/metal lateral junctions. The contribution is large even in junctions with long graphenes due to linear energy dispersion near the Fermi level. Second, the maximum conductance ($\Gamma = e^2/h$) state, identified to be an effective Dirac point, shifts with graphene length and band mixing between graphene and metal electrodes. The spin-dependent shift in DPs is effective when the contacts between graphene and electrodes are metallic. A combined effect of the matching/mismatching of conductive channels and a change in the local electronic states near DPs may result in a high MR ratio in some transition metal alloys, when disorder is sufficiently small in junctions. The qualitative results are independent of junction structures.

ACKNOWLEDGMENTS

The work was partly supported by the Next Generation Super Computing Projects, NanoScience Program, MEXT, Japan, Grants-in-Aids for Scientific Research in the priority area "Spin current" from MEXT, Japan, and Elements Science and Technology Projects of MEXT, Japan.

REFERENCES

1. P. Grünberg, R. Schreiber, Y. Pang, M. B. Brodsky, and H. Sowers, *Phys. Rev. Lett.* **57**, 2442 (1986).
2. M. N. Baibich, J. M. Broto, A. Fert, F. Nguyen Van Dau, F. Petroff, P. Etienna, G. Creuzet, A. Friederich, and J. Chazelas, *Phys. Rev. Lett.* **61**, 2472 (1988).
3. T. Miyazaki and N. Tezuka, *J. Magn. Magn. Mater.* **139**, L231 (1995).
4. J. S. Moodera, L. R. Kinder, T. M. Wong, and R. Meservey, *Phys. Rev. Lett.* **74**, 3273 (1995).
5. J. Inoue, A. Oguri, and S. Maekawa, *J. Phys. Soc. Jpn.* **60**, 376 (1991).
6. J. M. MacLaren, X.-G. Zhang, W. H. Butler, and X. Wang, *Phys. Rev. B* **59**, 5470 (1999).
7. J. Mathon, and A. Umerski, *Phys. Rev. B* **60**, 1117 (1999).
8. K. S. Novoselov, A. K. Geim, S. V. Morozov, D. Jiang, M. I. Katsnelson, I. V. Grigorieva, S. V. Dubonos, and A. A. Firsov, *nature* **438**, 197 (2005).
9. A. H. Castro Neto, F. Guinea, N. M. R. Peres, K. S. Novoselov, and A. K. Geim, *Rev. Mod. Phys.* **81**, 109 (2009).
10. N. Tombros, C. Jozsa, M. Popinciuc, H. T. Jonkman, and B. J. van Wees, *nature* **448**, 571 (2007)
11. E. W. Hill, A. K. Geim, K. Novoselov, F. Schedin, and P. Blake, *IEEE Trans. Magn.* **42**, 2694 (2006).
12. M. Nishioka and A. M. Goldman, *Appl. Phys. Lett.* **90**, 252505 (2007).
13. M. Ohishi, M. Shiraishi, R. Nouchi, T. Nozaki, T. Shinjo, and Y. Suzuki, *Jpn. J. Appl. Phys.* **46** L605 (2007).
14. A.Yamamura, S. Honda, J. Inoue, and J. Itoh, *J. Magn. Soc. Jpn.* **34**, 34 (2010).
15. S. Honda, A. Yamamura, T. Hiraiwa, R. Sato, J. Inoue, and H. Itoh, *Phys. Rev. B* **82**, 033402 (2010).
16. W. Harrison, *Electronic structure and the properties of solids*, W. H. Freeman and Company (1980).
17. D. A. Papaconstantpoulos, *Handbook of the band structure of elemental solids* (Plenum Press, New York, 1986).
18. H. Itoh and J. Inoue, *J. Magn. Soc. Jpn.* **30**, 1 (2006).

Mater. Res. Soc. Symp. Proc. Vol. 1284 © 2011 Materials Research Society
DOI: 10.1557/opl.2011.647

Room Temperature Superparamagnetism observed in Foam-like Carbon Nanomaterials

Shunji Bandow, Hirohito Asano, Susumu Muraki, Takahiro Mizuno,
Makoto Jinno and Sumio Iijima

Department of Materials Science and Engineering, Meijo University,
1-501 Shiogamaguchi, Tenpaku, Nagoya 468-8502, Japan

ABSTRACT

Magnet-attractive carbon nanopowder can be produced by a pulsed Nd:YAG laser (10 Hz) vaporization of pure carbon in a few % of H_2 containing Ar gas at 1000°C. On the other hand, magnet-attractive nanopowder cannot be formed when vaporizing in pure Ar. As-grown carbon nanopowder includes a few to ten % of micron sized graphite flakes as the impurity. Removal of such flakes can be achieved by a centrifugal separation and the supernatant is checked by X-ray diffractometry (XRD) and transmission electron microscopy (TEM). Magnetization curve at 400 K is easy to saturate at low magnetic field of 10 kG, and no hysteresis is observed. This feature is explained by a superparamagnetism of finely dispersed ferro- or ferri-magnetic nanoparticles. Elementary analyses using electron energy loss spectroscopy (EELS) and atomic absorption spectroscopy (AAS) suggest that the observed strong magnetism should be an intrinsic carbon magnetism.

INTRODUCTION

All carbon nanomagnet was first developed by Rode et al. in 2004 by using high repetition rate (2-25 kHz) pulsed laser vaporization of carbon in pure Ar gas [1]. However, the strong magnetism was observable only at low temperatures (< ~90 K) and this magnetism was not stable in open air. Theoretically, strong carbon magnetism is associated with the zig-zag edge of graphene where the non bonding electron spins are localized in the narrow band near the Fermi level [2, 3]. The electron spins on such zig-zag edge have a tendency to orient their directions with a ferromagnetic manner and the finite number of magnetic moment will be appeared. On the other hand, the electrons on the armchair face do not indicate such magnetism. Generally, the non bonding electron spins are quite active, so that these spins immediately react with other chemical species such as O_2 and H_2O in air. This will result the non magnetic state.

Carbon is a light element and exists widely on the earth. In addition, the strong magnets are all composed of heavy rare earth elements. Hence the carbon magnet will be pioneering and environment-friendly material. In this study, a reproducible method for making the carbon nanomagnet is introduced, and the purified product is analyzed in detail by using TEM, XRD, SQUID (superconducting quantum interference device), EELS and AAS.

EXPERIMENTAL

Sample preparation

Pure carbon target for a pulsed Nd: YAG laser vaporization was produced by pressing the carbon powder (a Kojundo Kagaku, purity 99.7 % with the grain size ~5 μm) in a 10 mm diameter cylinder at 350 kg/cm^2 for 30 min. Then the pelletized carbon powder was carefully removed from the cylinder and treated by using non-magnetic tweezers. In order to get well-formed cylindrical target, it was necessary to use the carbon powder with the grain size less than 10 μm. Carbon target (10 mm in diameter and 5 mm in thick) thus prepared was mounted on the target holder made by high purity carbon and was put at the center of the quartz tube (25 mm in diameter). This quartz tube was set in the electric tube-furnace. Then the target space was evacuated by an oil free scroll pump. After reaching the base pressure, the quartz tube was filled with 3 % of H_2 containing Ar gas, and a continuous flow with 100 sccm at 1 atm was made in the tube. Next the furnace temperature was elevated to 1000°C and the laser vaporization of the target was carried out for 30 min at the power density of 3~4 J/cm^2. A few mg of product can be gathered from water cooled sample collector located at the downstream side. In addition, the laser vaporization was carried out in pure Ar flux and this sample was used for the control experiment.

Purification

As-grown product was dispersed in ethanol by using ultrasonic agitation (a Cosmo Bio, Nanoraputor NR-350). Then the colloidal solution was centrifuged at 500 G for the product prepared in H_2 containing condition and at 5000 G for pure Ar. Upper 2/3 of liquid was collected as supernatant and lower 1/3 was collected as sediments. Then the ethanol was dried and the remained powder was collected for further characterization.

Structural, magnetic and impurity analyses

Structural features were examined by both transmission electron microscope (a JEOL, JEM2010F) equipped with an EELS (a Gatan, GIF-678) and powder X-ray diffractometer (a Rigaku, Ultima-IV). Magnetization curve and temperature dependence of the magnetic susceptibility were measured by a SQIUD susceptometer (a Quantum Design, MPMS-7). Iron contents in the as-grown sample and the purified one were analyzed by an AAS (a Hitachi, Z-2310). The sample for AAS was prepared by dispersing 0.5 mg of product to 25 ml ethanol.

RESULTS and DISCUSSION

Structural characterization

TEM images of both supernatants prepared in H_2 containing condition (CNF-H_2-spnt) and in pure Ar (CNF-Ar-spnt) are shown in figure 1 together with a schematic of proposed structure and demonstrations of magnet-attractiveness. From these TEM images, one can find that the structures of both products closely resemble the foam. Hence we categorize these materials as carbon nanofoam (CNF). Such structural features are similar to the observation by Rode et al [1]. Schematic in the center column represents randomly connected graphene clusters, and these clusters are sterically overlapped. Such sterical overlap may protect zig-zag edge from air exposure. It should be emphasized here that even though both products have almost the same structural feature, magnet-attractiveness is quite different. CNF-H$_2$-spnt is strongly magnet-attractive, but the other is not.

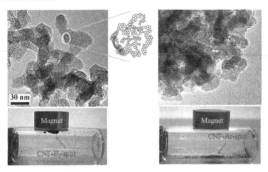

Figure 1. TEM images and demonstrations of manget-attractiveness for CNF-H$_2$-spnt (left) and CNF-Ar-spnt (right). Proposed structure of CNF (carbon nanofoam) is illustrated in the center column. Individual CNF particles agglutinate through a part of graphene layer and each particle size is around 20 nm.

XRD patterns of as-grown CNF-H$_2$, CNF-H$_2$-sed (sediments part) and CNF-H$_2$-spnt are indicated in figure 2. Narrow diffraction peak at $2\theta = 26.6°$ is associated with the diffraction from (002) of graphite and they are classified into impurity graphite flakes. This narrow peak can be seen only in as-grown and sediments of CNF-H$_2$. Characteristic of supernatant sample is

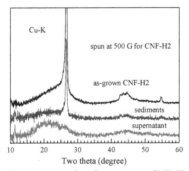

Figure 2. X-ray diffraction patterns taken for as-grown CNF-H$_2$ (top), CNF-H$_2$-sed (middle) and CNF-H$_2$-spnt (bottom). Narrow diffraction peaks at $2\theta = 26.6°$ detected for as-grown and sediments are from impurity graphite flakes. Hence pure CNF gives XRD pattern indicated at the bottom.

a broadened diffraction profile at around $2\theta \sim 21°$. This diffraction profile can be explained by (002) diffraction associated with a set of faced graphene layers (double layer diffraction) [4]. Such double layered part can be seen in TEM image indicated in figure 3. According to the diffraction theory, diffraction profile I is represented by a following equation:

$$I = a^2 \frac{\sin^2 a Q}{(a Q / 2)^2} \cdot \frac{\sin^2(N\,d_{002}\,Q / 2)}{\sin^2(d_{002}\,Q / 2)}. \quad (1)$$

Here, $a = 0.182$ nm (layer thickness for X-ray scattering), Q is the reciprocal of the plane spacing represented by $\dfrac{2\pi}{\lambda / 2\sin\theta}$ (λ is the wavelength of X-ray), N is the number of layers and d_{002} is the spacing of (002) plane. Analytical results of X-ray profiles are in figure 4. Fitting results clearly indicate that the (002) spacing is wider for CNF-Ar-spnt than that of CNF-H_2-spnt, and that the double layer part is necessary to explane the broadened peak profile.

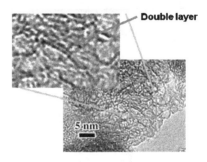

Figure 3. TEM images of CNF. A set of faced graphene layers forms double layered graphite.

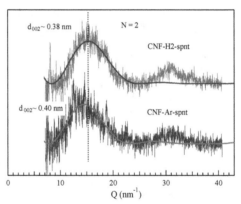

Figure 4. Analyses of X-ray profiles associated with (002) plane for supernatant samples. Thick solid lines are the fitting results using equation (1), which clearly indicate that the profiles are well fit by double layer diffraction ($N = 2$).

Impurity analyses

TEM-EELS analyses indicated that no Fe was detected for both CNF-H_2-spnt and CNF-Ar-spnt. On the other hand, a trace of Fe in CNF with the concentration of a few hundreds ppm

(~200-600 ppm) was detected for each sample by AAS. This contradiction can be easily understood when we consider the detection limits of these measurements. That is, EELS is in the order of 0.1 % (1000 ppm) and AAS is in 10 ppb order.

Magnetic features

Magnetization curves (MH curves) at 4.2 and 400 K are, respectively, indicated in figures 5 (a) and (b). Surprisingly, saturation of M can be seen even at 400 K for CNF-H_2-spnt in the magnetic field less than 10000 G. If the magnetism is associated with conventional paramagnetic spins ($S = 1/2$ etc.), saturation of M cannot be observed up to at least 10^7 G at 400 K. Thick solid curve on CNF-H_2-spnt in figure 5 (b) is a fitting result using Langevin function with the magnetic moment ~9,000μ_B. Hysteresis was started to observe below ~30 K. From such easy saturable M and hysteresis, it is necessary to consider that CNF-H_2-spnt should have tiny ferro- or ferri-magnetic domains, and the direction of magnetic moment will be flipped by thermal agitation. Such characteristic is typical for the superparamagnetism.

Magnitude of the saturation magnetization (M_s) was ~0.5 emu·G/g at 400 K. Here we calculate the magnitude of M_s on the assumption that the observed M_s simply originates in the Fe impurity that was detected by AAS with the concentration 200-600 ppm. Since the M_s for pure Fe is ~200 emu·G/g, magnitude of impurity magnetism is estimated to be 0.04 ~ 0.1 emu·G/g. These values are 10 to 5 times lower than the observed M_s. This means that the observed MH curves for CNF-H_2-spnt cannot be explained by the impurity magnetism. In addition, most important fact is that CNF-Ar-spnt did not indicate such MH curve even the same magnitude of Fe detected.

Temperature dependence of the magnetic susceptibility χ is shown in figure 5 (c). Blocking of the flipping motion of the magnetic moment of superparamagnetism can be seen below ~30 K for CNF-H_2-spnt. This means that the hysteresis observed below ~30 K in MH curve is purely the blocking effect on the thermally fluctuating magnetic moments. On the other hand, no remarkable temperature dependence was observed for CNF-Ar-spnt, which also suggests that the observed strong magnetism of CNF-H_2-spnt is intrinsic carbon magnetism.

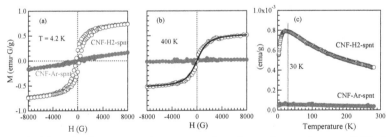

Figure 5. Magnetization curves and temperature dependence of magnetic susceptibility for CNF-H_2-spnt (open circles) and CNF-Ar-spnt (closed circles). MH curves at 4.2 and 400 K are, respectively, shown in (a) and (b). Temperature dependence of χ measured at 400 G is in (c). Thick solid line in (b) represents the fitting result using the Langevin function.

Next we consider the spin concentration of CNF-H_2-spnt on the hypothesis that the superparamagnetic spins are on the carbon atoms. When individual carbon atoms have 1 Bohr magneton (1 μ_B), M_s for 1 gram of carbon reaches 465 emu·G/g. Since our observation of M_s at 4.2 K was ~0.8 emu·G/g (see figure 5), it can be concluded that only ~0.2 % of carbon atoms have unpaired spin (~0.002 μ_B/carbon); in other words, 1 unpaired spin per ~500 carbon atoms. Size of each CNF particle was around 20 nm (see figure caption of figure 1). This means that the number of carbon atoms in a particle can be roughly estimated at ~3.6×10^5. Curve fitting of MH indicated a large magnetic moment reaching ~9,000μ_B. This roughly indicates that at least 10^4 carbon atoms are necessary to create such magnetic moment. In a CNF particle with 20 nm in diameter, the order of 10^5 carbon atoms are included. Hence we can find that about 1/10 CNF particles have such large magnetic moments [5].

CONCLUSIONS

Superparamagnetic foam-like carbon nanoparticles (carbon nanofoam, CNF) can be obtained by the pulsed laser vaporization of pure carbon in H_2 containing Ar gas at 1000°C. On the other hand, no such particle was obtained when the vaporization was carried out in pure Ar gas. Removal of impurity graphite flakes from the sample was achieved by the centrifugation using colloidally dispersed sample in ethanol. XRD profile of purified CNF particles was fit by the diffraction associated with a set of faced graphene layers (double layer diffraction). According to the magnetic measurements, it was found that the magnetization curve did not indicate hysteresis but easily saturated by ~10 kG at 400 K. In order to analyze this feature, we applied the Langevin function to fit the MH curve, and it was found that the thermally fluctuating large magnetic moment reaching ~$10^4\mu_B$ is necessary to explain this magnetism. Origin of this magnetism is not clear now, but it was not from spurious impurity origin.

ACKNOWLEDGMENTS

This work is partly supported by the Grants-in-Aid for Scientific Research (B) No. 22310070 of the Ministry of Education, Culture, Sports, Science and Technology (MEXT) from 2010 to 2012.

REFERENCES

1. A.V. Rode, E.G. Gamaly, A.G. Christy, J.G. Fitz Gerald, S.T. Hyde, R.G. Elliman, B. Luther-Davies, A.I. Veinger, J. Androulakis and J. Giapintzakis, *Phys. Rev.* **B70**, 054407 (2004).
2. M. Fujita, K. Wakabayashi, K. Nakada and K. Kusakabe, *J. Phys. Soc. Jpn.* **65**, 1920 (1996)
3. K. Nakada, M. Fujita, G. Dresselhaus and M. Dresselhaus, *Phys. Rev.* **B54**, 17954 (1996).
4. S. Bandow, F. Kokai, K. Takahashi, M. Yudasaka, L.C. Qin and S. Iijima, *Chem. Phys. Lett.* **321** 514 (2000).
5. H. Asano, S. Muraki, H. Endo, S. Bandow and S. Iijima, *J. Phys. Condens. Matter (Special issue: Carbon and related nanomaterials)* **22**, 334209 (2010).

Structural Characterization

Mater. Res. Soc. Symp. Proc. Vol. 1284 © 2011 Materials Research Society
DOI: 10.1557/opl.2011.221

In-situ Observations of Restructuring Carbon Nanotubes
via Low-voltage Aberration-corrected Transmission Electron Microscopy

Felix Börrnert[1], Alicja Bachmatiuk[1], Sandeep Gorantla[1], Jamie H. Warner[2], Bernd Büchner[1], and Mark H. Rümmeli[1,3]
[1]Leibniz-Institut für Festkörper- und Werkstoffforschung Dresden e. V., PF 270116, 01171 Dresden, Germany
[2]University of Oxford, Parks Road, Oxford OX13PH, United Kingdom
[3]Technische Universität Dresden, 01062 Dresden, Germany

ABSTRACT

The molecular structure and dynamics of carbon nanostructures is much discussed throughout the literature, mostly from the theoretical side because of a lack of suitable experimental techniques to adequately engage the problem. A technique that has recently become available is low-voltage aberration-corrected transmission electron microscopy. It is a valuable tool with which to directly observe the atomic structure and dynamics of the specimen *in situ*. Time series aberration-corrected low-voltage transmission electron microscopy is used to study the dynamics of single-wall carbon nanotubes *in situ*. We confirm experimentally previous theoretical predictions for the agglomeration of adatoms forming protrusions and subsequent removal. A model is proposed how lattice reconstruction sites spread. In addition, the complete healing of a multi-vacancy consisting of *ca.* 20 missing atoms in a nanotube wall is followed.

INTRODUCTION

Single-walled carbon nanotubes (SWNTs) are promising for application in future electronic devices [1]. Defects in the nanotubes can alter their electronic properties and are therefore studied intensely [2]. The behavior of defects in SWNTs is much discussed, but in most of the cases theoretically because of the special difficulty in handling nano-sized structures experimentally [3]. The behavior of sp^2 carbon under electron irradiation is of particular interest because of the possibilities that electron microscopes offer in the characterization and manipulation of this material. Smith and Luzzi calculate the carbon sp^2 lattice to be stable against knock-on damage under electron irradiation up to electron energies of 86 keV [4]. The lattice is most susceptible to the electron beam when the incident beam is normal to the lattice plane. Graphene is a perfect system to test this. Meyer and coworkers found that a clean graphene lattice is not affected by high-dose 80 keV electron irradiation but is damaged by lower-dose 100 keV electrons [5]. SWNTs as rolled-up graphene show a diameter dependent stability. Warner et al. demonstrated that small diameter nanotubes are destroyed more quickly than larger ones [6]. Also, they show contamination on the lattice leads to rapid destruction in the electron beam. In electronic devices hetero-junctions are most desired. Molecular junctions consisting of nanotubes and organic molecules or even single carbon chains have been proposed [7,8]. A convenient way to produce junctions is to have SWNTs with different chirality connected. These could be produced by locally induced material loss and subsequent lattice reconstruction as modeled by Ding et al. [9]. Rodriguez-Manzo and Banhart showed local material loss by drilling a hole into the nanotube wall with an electron beam [10].

Here, we directly image the dynamics of the SWNT structure *in situ* by means of time series aberration-corrected low-voltage transmission electron microscopy (TEM) [11]. We propose an explanation why smaller carbon nanotubes are destroyed more quickly under the electron beam than larger ones.

EXPERIMENT

The SWNTs employed were produced by a laser ablation route and have a mean diameter of approximately 1.5 nm [12]. For imaging, the sample was drop coated onto standard lacey carbon TEM grids. A FEI Titan[3] 80–300 transmission electron microscope with a CEOS aberration corrector for the objective lens, operating at an acceleration voltage of 80 kV, equipped with a Gatan UltraScan 1000 camera was used. All studies were conducted at room temperature. During a time series, an image was taken every 5 s with an acquisition time of 0.5 s. The contrast of the micrographs was enhanced through Fourier filtering by cutting frequencies beyond the information limit of the microscope. The chirality of the SWNTs was determined by analyzing the Fourier-transformed TEM images and measuring the corrected SWNT diameter [13]. Simulations of the imaging process were obtained using JEMS electron microscopy software [14]. For comparison, noise was added to the simulated images and the same Fourier filtering applied as for the TEM images.

RESULTS & DISCUSSION

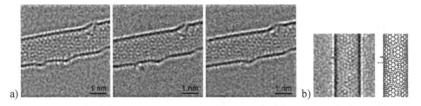

Figure 1. a) *In situ* TEM image series of a SWNT with a hump-like defect healing *via* restructuring into a protrusion. B) Simulation of the imaging process of a SWNT with three-atom protrusions.

In Figure 1, a micrograph series shows a SWNT under the electron beam. The time elapsed between each frame is 5 s. The circle highlights a region where a hump-like lattice defect restructures and subsequently vanishes. The large feature outside the tube in the second frame must be static for at least the camera's acquisition time. That is, the molecular structure giving this contrast must be bound to this lattice position. From simulations of the imaging process we conclude that the protrusion consists of a three atom chain. The smaller feature beneath the large one can be produced by another three-atom structure as seen in figure 1b). Molecular models with more or less atoms did not result in fitting simulated micrographs. The formation of three-atom protrusions from defects agrees well with calculations from Tsetseris and Pantelides who found that adatoms on a sp^2 lattice tend to agglomerate and form very mobile three-atom chains

[15]. We see an initially static protrusion that vanishes in the last frame. Most probably, the protrusion is trapped by the remaining defective lattice it just evolved from. Along with the reconstruction of the tube wall it vanishes from our images. Either the protrusion ejects into the vacuum or it becomes mobile and diffuses on the tube surface. In both cases the additional material is removed from the former defect site.

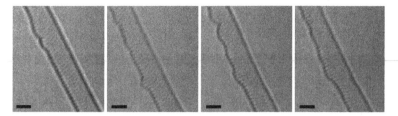

Figure 2. *In situ* TEM image series showing the spreading of a defective region in a SWNT lattice. The bar indicates 1 nm.

Figure 2 shows the development of a defect in the SWNT's left side. The shown frames are from a 11.5 minutes sequence. With time, the defective region spreads along the tube axis. The right side of the nanotube remains unchanged. From our experience, a non-defective nanotube is stable under an 80 kV electron beam. This is in agreement with the calculations of Smith and Luzzi as well as with the observations of Meyer *et al.* in the case of graphene [4,5]. According to Smith and Luzzi, the normal incident beam has the lowest threshold for knock-on damage, *i. e.*, one would expect the nanotube to be most susceptible to the damage at the front and back wall. We always observe the nanotubes to start the restructuring of the lattice at the side walls. This indicates that the energy for the restructuring comes not from knock-on interaction but more likely from other beam interactions with the electronic environment. Defects in the wall seem to be the precondition for any restructuring under the electron beam. Warner *et al.* show that dirt on the lattice considerably quickens the destruction of nanotubes [6]. Another observation we want to emphasize is that the defective regions spread along the tube axis. They do not move around the tube. We suggest two possible reasons for this behavior; either the interaction cross-section of the driving energy from the beam is considerably less if the beam falls perpendicularly onto the lattice, or the curvature of the lattice in the tube lifts the degeneracy in the local vibration amplitude of the carbon atoms. This would cause a lower probability of a bond site jump in a Stone-Wales mechanism around the tube than parallel to the tube axis. This also makes the lattice area involved in a defect reconstruction process considerably smaller than in a planar lattice. This might be an explanation of the preferential destruction of smaller tubes with a stronger curvature as observed by Warner *et al.* [6]. Generally it is assumed that the temperature of a SWNT remains stable under the electron beam due to the high heat conductivity and the small interaction cross-section of the carbon lattice with the beam. Nevertheless, the local properties are altered by defects, thus we do not exclude heat as the driving energy for the lattice reconstruction.

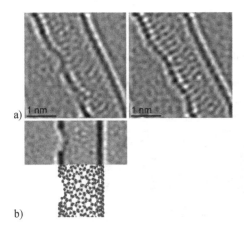

Figure 3. a) TEM images of a hole closing. The time elapsed between the two frames is 20 s. b) Simulation of the imaging process to fit the hole size.

The micrographs in figure 3 show two holes in the tube wall that are healed within 20 s. The simulations of the imaging process shown in 3b) let us estimate the size of the bigger observed hole of about 20 missing atoms. We find no evidence that the hole is rotated or moved elsewhere. Also, it is clearly seen that the tube diameter at the former hole positions is decreased. If the reconstruction mechanism subsequently builds a perfect sp^2 lattice again, we have produced a junction between nanotubes of different diameter. A controlled production of defined holes already has been demonstrated by Rodriguez-Manzo and Banhart [11].

CONCLUSIONS

In this work, we show the self-healing dynamics of single-walled carbon nanotubes *via* aberration-corrected low-voltage transmission electron microscopy. The formation of three-atom protrusions at the nanotube wall is observed, confirming previous theoretical predictions. Moreover, we experimentally demonstrate the complete healing of the tube wall after material removal. This confirms previous models. Explanations for the preferred destruction of smaller-diameter tubes previously reported are discussed. In addition, the closing of multivacancies of up to approximately 20 missing atoms is presented.

ACKNOWLEDGMENTS

FB acknowledges the DFG (RU 1540/8-1), AB the A.-v.-Humboldt Stiftung and the BMBF, SG the ''Pakt für Forschung und Innovation'', and MHR the EU (ECEMP) and the Freistaat Sachsen.

REFERENCES

1. P. Avouris, Acc. Chem. Res. **35**, 1026 (2002).
2. J.-C. Charlier, Acc. Chem. Res. **35**, 1063 (2002).
3. A. V. Krasheninnikov and K. Nordlund, J. Appl. Phys. **107**, 071301 (2010).
4. B. W. Smith and D. E. Luzzi, J. Appl. Phys. **90**, 3509 (2001).
5. J. C. Meyer, A. Chuvilin, and U. Kaiser in *Materials Science*, edited by W. Grogger, F. Hofer, and P. Pölt, (*MC2009*, Verlag der Technischen Universität Graz, Graz, 2009) pp. 347-348.
6. J. H. Warner, F. Schäffel, G. Zhong, M. H. Rümmeli, B. Büchner, J. Robertson, and G. A. D. Briggs, ACS Nano **3**, 1557 (2009).
7. M. del Valle, R. Gutiérrez, C. Tejedor, and G. Cuniberti, Nature Nanotechnol. **2**, 176 (2007).
8. F. Börrnert, C. Börrnert, S. Gorantla, X. Liu, A. Bachmatiuk, J.-O. Joswig, F. R. Wagner, F. Schäffel, J. H. Warner, R. Schönfelder, B. Rellinghaus, T. Gemming, J. Thomas, M. Knupfer, B. Büchner, and M. H. Rümmeli, Phys. Rev. B **81**, 085439 (2010).
9. F. Ding, K. Jiao, Y. Lin, and B. I. Yakobson, Nano Lett. **7**, 681 (2007).
10. J. A. Rodriguez-Manzo and F. Banhart, Nano Lett. **9**, 2285 (2009).
11. F. Börrnert, S. Gorantla, A. Bachmatiuk,, J. H. Warner, I. Ibrahim, J. Thomas, T. Gemming, J. Eckert, G. Cuniberti, B. Büchner, and M. H. Rümmeli, Phys. Rev. B **81**, 201401(R) (2010).
12. M. H. Rümmeli, C. Kramberger, M. Löffler, O. Jost, M. Bystrzejewski, A. Grüneis, T. Gemming, W. Pompe, B. Büchner, and T. Pichler, J. Phys. Chem. B **111**, 8234 (2007).
13. A. Hashimoto, K. Suenaga, A. Gloter, K. Urita, and S. Iijima, Nature (London) **430**, 870 (2004).
14. P. A. Stadelmann, Ultramicroscopy **21**, 131 (1987).
15. L. Tsetseris and S. T. Pantelides, Carbon **47**, 901 (2009).

Optical Probes

Mater. Res. Soc. Symp. Proc. Vol. 1284 © 2011 Materials Research Society
DOI: 10.1557/opl.2011.648

A fully automated remote controllable microwave-based synthesis setup for colloidal nanoparticles with integrated absorption and photoluminescence online analytics

Simon Einwächter [1,2], Michael Krüger[1,2]
[1]Freiburg Materials Research Centre, University of Freiburg, D-79104 Freiburg, Germany
[2]Institute for Microsystems Technology, University of Freiburg, D-79110 Freiburg, Germany

ABSTRACT

We present a fully automated microwave-based synthesis setup for colloidal nanoparticles. Integrated absorption and photoluminescence online analytics opens the possibility to monitor the growth of various nanoparticles at any stage of the reaction. Spectroscopic investigation within the first seconds of a reaction is accessible opening the possibility to detect potential critical size nuclei as a function of the reaction conditions. Beside the possibility to perform systematic mechanistic studies, this system allows a high degree of synthesis control leading to very good product reproducibility. In conjunction with an automated auto sampler unit systematic multiple reactions can be performed one after each other and compared. The setup is remote-controllable allowing worldwide online control accessibility over the synthesis setup including data processing, visualization and storage. The performance of the setup will be demonstrated by using the synthesis of CdSe nanocrystals as a model system and can be extended to the synthesis of various metallic and semiconducting nanoparticles.

INTRODUCTION

One major challenge for the synthesis of nanomaterials and nanoparticles is the reproducibility and the up scalability of the synthesis in terms of controlling the size, morphology, surface constitution and crystallinity of the resulting particles which often determines specific chemical and physical properties. In chemical laboratories colloidal nanoparticles are often synthesized via classical routes using standard flasks and equipment such as heaters and stirrers etc. By even using the same synthesis protocol, two different people often achieve different results depending on their individual interpretation of the protocols and differences in physical and chemical parameters influencing the reaction which are usually not described and well documented such as, heating rate, stirring velocity, size of the reaction vessel, chemical impurities etc. Some of the critical parameters for the synthesis of CdSe quantum dots using the manual colloidal hot injection synthesis method are described in Ref. [1]. Microwave based approaches as well as microfluidic reactor based approaches for the synthesis of nanoparticles have been explored for both, improving the control over the reaction for enhancing reproducibility and for up scaling syntheses which is especially important for the development of applications based on nanoparticles and nanoparticle hybrid materials. While microwave based synthesis approaches are usually performed in commercially available instruments the microfluidic approach needs more engineering and designing skills. Strauss and Co workers for example used a microwave reactor not only as reproducible heating source but also for enhancing the reaction rate of CdS nanoparticles due to the selective microwave absorption of the precursor resulting in lower reaction temperatures [2]. Chan et al. built for example a microfluidic reactor for a controlled growth of CdSe nanocrystals [3] and a more advanced automated version called "WANDA" has recently been developed for reproducible high-throughput synthesis approaches for colloidal nanocrystals [4], which represents a whole workstation integrated into a glove box system for automated nanomaterial discovery and analysis.

Here we present a novel synthesis setup based on a commercially available microwave synthesis reactor with integrated absorption or photoluminescence spectroscopic detection systems for the synthesis of nanoparticles using CdSe quantum dots as a model system for evaluating and optimizing the synthesis setup.

EXPERIMENTAL RESULTS AND DISCUSSION

Experimental setup

The set-up consist out of a microwave reactor (Discover S-Class, CEM Corporation, USA) with a radiation frequency of 2.4 GHz for performing reactions at pressures up to 17 bars (Fig. 1 a). Different tube sizes are available as reaction vessels. For the integration of absorption and photoluminescence immersion probes (All-Quartz Immersion Probe with 2 mm optical path length from Hellma GmbH & Co. KG, Germany) a dedicated glass vessel was constructed (Fig. 1b) allowing the integration of the metal free fiber optical coupled sensors directly into the reaction solution. The fiber optical sensors were coupled to an absorption spectrometer (Tidas 100, J & M - Analytik AG, Germany) and a fluorescence spectrometer (Fluorolog 3-22, Horiba Yvon, Japan) respectively to realize online spectroscopic investigations (Fig. 1c). Alternatively, a fiber optical system can be installed outside the reaction tube in order to detect photoluminescence in a reflection mode. Special fiber optics (LOPTEK Glasfasertechnik GmbH & Co. KG, Germany) were installed in close contact to the reaction tube (Fig 1 d).

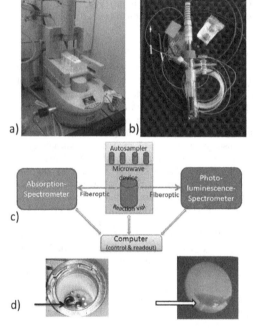

Fig.1: a) Commercially available computer controlled laboratory microwave system with integrated video camera and auto sampler. b) Dedicated reaction vial allowing the integration of an absorption or fluorescence immersion probe head to perform syntheses under nitrogen atmosphere. c) Schematics of the remote controllable microwave synthesis reactor with integrated fiber optical based online absorption and fluorescence detection possibilities. d) External fluorescence probe for reflective online fluorescence measurements integrated into the microwave reactor chamber (top view left and side view right (video camera image)). The fiber optical probe (marked with an arrow) is installed nearby the reaction vessel.

Synthesis of CdSe quantum dots

For each reaction, typically 1.21 g (5 mmol) hexadecanol, 40.5 mg (50 μmol) cadmium laureate and 50 μl of 1 M solution of Selene in trioctylphosphine was mixed and capped in a vial under nitrogen atmosphere in a glove box. In order to homogenize the mixture, the sample was heated in the microwave to 65°C for 3 minutes at maximum stirring speed and 90 W microwave power. After this the temperature was increased to typically 160°C (at 300 W microwave power reached within 2 minutes) to generate CdSe quantum dots. A color change of the reaction solution indicates the start of quantum dot formation. The reaction can be followed by online absorption and/or photoluminescence spectroscopy to monitor the reaction progress.

Online absorption spectroscopy

Online absorption spectroscopy can be performed using an all-quartz dip in immersion probe (2 mm optical path length, Hellma GmbH & Co. KG, Germany) fiber optically coupled to an absorption spectrometer. About 10 spectra per second can be recorded easily allowing the detection of fast changes during the reaction especially in the beginning of the reaction. An overlay of a series of absorption spectra recorded during a CdSe quantum dot synthesis is displayed in Fig.2.

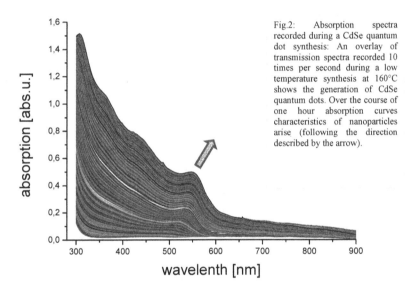

Fig.2: Absorption spectra recorded during a CdSe quantum dot synthesis: An overlay of transmission spectra recorded 10 times per second during a low temperature synthesis at 160°C shows the generation of CdSe quantum dots. Over the course of one hour absorption curves characteristics of nanoparticles arise (following the direction described by the arrow).

By using immersion probes for online-spectroscopy the reaction conditions are restricted to a maximum probe temperature of about 160°C and over pressures up to 6 bar, without a durable fixation of the probe. An external detection of the absorption is envisioned to allow full automation using the auto sampler unit and widening the range of synthesis approaches (e.g. for higher temperature and higher pressure syntheses). Therefore evaluations and tests are ongoing to establish an optical system that detects the absorption in a non-contact mode enabling the auto sampler to exchange the high pressure reaction vials without any manual interference.

Online photoluminescence spectroscopy

Besides the possibility to use dip in probes similar as described for the online absorption spectroscopy an external non-contact detection is possible. For photoluminescence spectroscopy, a fiber bundle with the excitation channel concentrically surrounding the emission light channel is used to measure photoluminescence in reflection mode: The materials of this fiber optic have been tested for suitability to be used in the microwave. The tip of the fiber and the tubing consist out of non-blackened PTFE, and the silica fiber cores and claddings are microwave-transparent. The epoxy resin withstands the temperatures caused by its moderate heat up in the microwave field. This fiber optic is directed through the modified pressure release chamber of the microwave into the cavity. The reaction solution is screened in a non-contact mode across the glass vials. Therefore standard capped commercially vials can be used in combination with the pressure controlling system that handles up to 17 bar as well as the auto sampler unit. In Fig. 2 a series of photoluminescence peaks is recorded during the synthesis of CdSe quantum dots. The shift of the signal towards longer enission wavelengths indicates the growth of CdSe nanoparticles. The intensity of the peaks initially raises and decreases after reaching a maximum at around 570 nm.

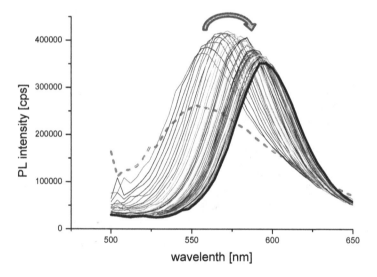

Fig.3: Photoluminescence spectra recorded during a CdSe quantum dot synthesis. An overlay of photoluminescence spectra is shown that was recorded every 15 seconds during a low temperature synthesis approach of CdSe quantum dots over a total time of 15 min. The first spectrum taken is highlighted with a dashed line and the final spectrum with a thicker line respectively. The arrow indicates the temporal evolution of the spectra.

Here, the present limitations for the reaction conditions are due to temperature and concentration dependent fluorescence quenching resulting in a weakening of the signal at higher reaction temperatures and concentrations. In our model system we could detect the fluorescence signal up to 220°C– 260°C in concentrated reaction solutions. Since microwave synthesis usually favors lower reaction temperatures [2] this restriction seems to be not limiting that much. Furthermore, an improvement of the detection system leading to higher sensitivity is possible by focusing the emission light with additional optical components (like lenses or mirrors) and is under investigation in order to allow online photoluminescence spectroscopy at higher reaction temperatures with a reasonable time resolution.

Remote-control of the synthesis

The reaction system is remote-controllable via a java applet included in the open source software ultraVNC.The applet runs in most internet browsers and on most operating systems (Fig. 4). So the scientist can start and monitor reactions and interact during syntheses from outside of the laboratory. Further development of this technology is currently being made within the project "BW-eLabs". There a whole software platform is created for the integration and the management of remote controllable equipment including data storage and presentation. A substantial characteristic of BW-eLabs is that the corresponding data and documents are examined along their entire life cycle from synthesis, material characterization to the integration into applications and embedded into the entire process chain of experimental environments [5].

Fig.4: Remote control scenario. The system can be controlled from any computer with internet access using a browser and installed Java runtime environment to run a browser applet based on the open source software ultraVNC. This scenario shows two computers standing in an office. With the left one the microwave synthesis reactor in the laboratory can be controlled and running reactions can be monitored. The right one controls and monitors the online analytical spectroscopy.

DISCUSSION AND OUTLOOK

A reaction system based on a remote controllable laboratory microwave oven and integrated online absorption and photoluminescence spectroscopy has been build up and tested for the generation of CdSe quantum dots. Although a decent grade of automation has been achieved, further improvement of non-contact online absorption spectroscopy is desired and is currently being developed.

A fast heating rate simulates hot injection methods where the nucleation and growth of nanoparticles can be separated well enough and thereby helps to simplify synthesis approaches and maintain reproducibility for the synthesis of nanoparticles. The high reproducibility of syntheses performed with this system as well as the possibility to monitor the reaction by online spectroscopy with high time resolution enables a variety of fundamental investigations on growth processes occurring in nanocrystals. Furthermore, closely defined protocols for microwave syntheses can be published and reproduced easily (without further optimization effort). The synthesis set-up is also aimed to enhance the development of new synthesis approaches for novel nanomaterials. The integration of the synthesis setup into a broader platform connecting existing infrastructures such as e.g. digital libraries, decentralized tools and repositories under Open Access-Policy is currently undertaken.

ACKNOWLEDGMENTS

We acknowledge Dr. A. Liehr, K. Zimmermann and R. Würdemann for fruitful discussions and the Ministry of Science and Education of Baden Württemberg (Germany) for funding the currently running project "BW-eLabs".

REFERENCES

1. Y. Yuan, F. S. Riehle, H. Gu, R. Thomann, G. Urban, M. Krüger, *J. Nanosci. Nanotechnol.* **10**, 6041-6045, (2010).
2. A. L. Washington, G. F. Strouse, *Chem. Mater.* **21**, 3586-3592, (2009).
3. E. M. Chan, R. A. Mathies, P. A. Alivisatos, *Nano Letters* **3 (2)**, 199-201, (2003).
4. E. M. Chan, C. Xu, A. W. Mao, G. Han, J. S. Owen, B. E. Cohen, D. J. Milliron, *Nano Letters*, **10**, 1874-1885, (2010).
5. S. Jeschke, B. Burr, J.-U. Hahn, L. Helmes, W. Kriha, M. Krüger, A.W. Liehr, W. Osten, O. Pfeiffer, T. Richter, G. Schneider, W. Stephan, K.-H. Weber, 10th ACIS International Conference on Software Engineering, Artificial Intelligences, Networking and Parallel/Distributed Computing, *IEEE Computer Society*, 47-52, (2009).

Electronic, Optical, and Magnetic Properties of Carbon Nanomaterials II

Mater. Res. Soc. Symp. Proc. Vol. 1284 © 2011 Materials Research Society
DOI: 10.1557/opl.2011.222

Band Gap Opening of Graphene after UV/Ozone and Oxygen Plasma Treatments

Adrianus I. Aria[1], Adi W. Gani[2] and Morteza Gharib[1]
[1]Graduate Aeronautical Laboratories, California Institute of Technology, Pasadena, California 91125, USA
[2]Electrical Engineering, California Institute of Technology, Pasadena, California 91125, USA

ABSTRACT

Graphene grown by Chemical Vapor Deposition (CVD) on nickel subsrate is oxidized by means of oxygen plasma and UV/Ozone treatments to introduce bandgap opening in graphene. The degree of band gap opening is proportional to the degree of oxidation on the graphene. This result is analyzed and confirmed by Scanning Tunnelling Microscopy/Spectroscopy and Raman spectroscopy measurements. Compared to conventional wet-oxidation methods, oxygen plasma and UV/Ozone treatments do not require harsh chemicals to perform, allow faster oxidation rates, and enable site-specific oxidation. These features make oxygen plasma and UV/Ozone treatments ideal candidates to be implemented in high-throughput fabrication of graphene-based microelectronics.

INTRODUCTION

Graphene is a lattice of honeycomb-arranged carbon atoms tightly joined by sp^2 bonds [1]. Graphene draws a lot of research interests due mostly to its unique band structures and excellent electrical properties [2]. Graphene has a superior carrier mobility and lowest resistivity than any other materials. These features enable graphene to be a promising future candidate for component of integrated circuit. However, graphene is a semimetal and zero bandgap material. This puts limitation on controlling and switching off electrical conductivity of graphene by means of gate electrode [3]. In order to harvest the maximum potential of graphene as a promising future electronic material, a number of approaches are currently done to create a bandgap opening in graphene. When single layer graphene is cut into ribbons of nanometers wide, a bandgap appears as a result of quantum confinement [4-6]. Another approach to introduce bandgap in grapheen is to apply electric field normal to a graphene plane [3]. Another method is to break the lattice symmetry of pristine graphene by introducing "dopant" atoms or functional groups, such as bismuth, antimony, gold [7], NO2 [8], and oxygen [9]. Among many dopant species that have been demonstrated to introduce band opening in graphene, oxygen has its own virtue for further refinements, because its chemical interaction with organic molecules have been well-understood.

Current method of preparing graphene oxide stems mainly from established wet-chemical methods of preparing graphite oxide, such as the Hummer's method [10]. These methods generally use graphite flakes as the starting material, which is oxidized in the presence of strong acids and oxidizing agents. The resulting graphite oxide consists of layered structure of strongly-hydrophilic graphene oxide, that enables intercalation of water molecules between the graphene oxide layers. A sonication or stirring will exfoliate the graphene oxide, producing aqueous colloidal suspensions of graphene oxide [11], which is then washed, filtered, and dried to obtain dry graphene oxide flakes. The wet-oxidation methods all employ harsh chemicals and significant amount of time to finish. For instance, the commonly-used modified Hummer's method [12-14] typically uses sulfuric acid and potassium permanganate during the oxidation process, and usually takes hours to complete. In addition, wet oxidation methods lack site-specificity, since it can only be utilized to oxidize graphene in bulk. As a result, these wet processes may be incompatible to be implemented in graphene-based integrated circuit fabrication process.

We report a dry oxidation method using oxygen plasma or UV/Ozone treatment to prepare graphene oxide. In contrast with the established wet methods of preparing graphene oxide, dry oxidation method reported here does not use harsh chemicals at all, and require significantly less oxidation time. In addition, O2 plasma and UV/Ozone treatment processes have been well-integrated in current silicon-based integrated circuit mass-manufacturing process, and it allows site-specific oxidation through conventional photolithrographic masking steps.

EXPERIMENTAL DETAILS

Graphene samples used in this study were grown by chemical vapor deposition (CVD) with nickel catalyst on SiO2/Si substrate. For UV/ozone treatment, cleaned graphene samples were placed inside a UV/Ozone cleaner (Bioforce Nanosciences), and exposed for various time lengths at room temperature (UVO samples). Other sets of samples (O2P samples) were treated with oxygen plasma (Tepla M4L) for a couple of seconds with 20 Watts of RF power while chamber condition is maintained at 500 mTorr and 20 SCCM. The surface of the samples is analyzed with Raman Spectroscopy (Renishaw M1000). Current images of samples, tunneling I-V characteristics, and tunnelling differential conductance are obtained with scanning tunnelling microscope/spectroscope (Digital Instrument Nanoscope IIIa ECSTM). The scanning tip is made of Pt/Ir (Veeco, Inc). All experiments involving the STM/STS were done in room temperature and atmospheric pressure. Energy dispersive x-ray spectrometer (Oxford INCA Energy 300) was used to measure the atomic ratio of oxygen to carbon present in oxidized graphene.

DISCUSSIONS

The electronic properties of graphene treated with oxygen plasma and UV/ozone were probed by a scanning tunneling spectroscopy (STS). Figure 1 shows the tunneling current as a function of bias voltage for graphene treated with oxygen plasma and UV/ozone for various exposure times. As graphene undergoes a longer exposure to oxygen plasma and UV/ozone treatment, its tunneling current around the zero-bias region gradually decreases to zero, eventually creating a flat region where no tunneling current flows through. At this stage, additional oxygen plasma and UV/ozone exposure to graphene further widens the flat region. Even though both dry-oxidation treatments produce a similar effect on the I-V characteristic of oxidized graphene, oxygen plasma treated graphene achieves the I-V flattening effect in a much faster fashion than UV/ozone treated graphene. As shown in figure 1, a pronounced flattening of I-V characteristic of oxygen plasma treated graphene has been achieved after an exposure of just less than 30 seconds (O2P30s samples), whereas a pronounced flattening of I-V characteristic of UV/ozone treated graphene has been achieved after 30 minutes of exposure (UVO30m samples).

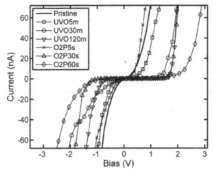

Figure 1. Effect of oxygen plasma and UV/ozone treatments on tunneling I-V characteristic of graphene as probed using a scanning tunneling spectroscopy (STS).

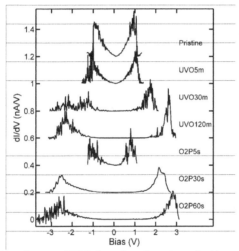

Figure 2. Differential conductance dI/dV plots of graphene treated with oxygen plasma and UV/ozone for various lengths of time. In both treatments, as graphene undergoes a longer duration of oxidation, the flat region around 0 Volt bias where its differential conductance is zero gets wider, indicating a larger band gap opening of oxidized graphene.

Figure 2 shows tunneling differential conductance dI/dV plots of oxygen plasma and UV/ozone treated graphene obtained by numerically taking the first derivative of the tunneling I-V characteristics data. As shown in figure 2, the local density of states (LDOS) of pristine graphene, indicated by its tunnelling differential conductance dI/dV, shows Dirac point where the local density of states of graphene is at zero minimum. As graphene is oxidized with oxygen plasma and UV/ozone treatments, the local density of states gradually starts to flatten out, creating a band gap opening. For instance, after 60 seconds of exposing graphene to oxygen plasma treatment (O2P60s samples), the differential conductance plot suggests the oxidized graphene has a band gap opening of 2.4 eV. Similarly, 120 minutes of UV/ozone exposure gives the oxidized graphene a band gap of 1.9 eV. The flattening of the local density of states suggests a transformation of electrical behavior of graphene from semi-metallic to semiconducting.

The degree of band gap opening as a function of oxidation time for both oxygen plasma and UV/ozone treatments is summarized in figure 3. It is worth noting that in both treatments the rate of band gap opening gradually decreases over lengths of oxidation time. A plateau band gap value of approximately 2.4 eV has been achieved by O2P60s. The decreasing rate of band gap opening can be explained by understanding the mechanism of oxidation of graphene. Sites with lattice defects are more prone to be functionalized with oxygenated functional groups than sites without any lattice defects. As graphene is exposed to oxygen plasma and UV/ozone treatments, oxygenated functional groups will quickly form at defect sites. As oxidation goes further, less defect sites remain, which makes it harder for reactive oxygen species to find defect sites and form additional oxygenated functional groups with graphene. Because the degree of band gap opening in graphene positively correlates to the atomic ratio of oxygen and carbon (figure 4), the rate of band gap opening eventually slows down.

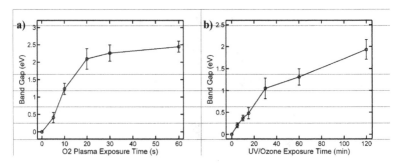

Figure 3. Band gap opening in graphene treated with **a)** oxygen plasma and **b)** UV/ozone for various exposure times.

Raman spectroscopy measurement was conducted to characterize the structural properties of graphenes before and after oxidation, which was also employed to verify their electronic properties. Raman spectra of graphenes before and after oxidation process can be seen in figure 5. In agreement with previously reported studies [15-17], pristine monolayer CVD graphene shows distinguishable D band (~1350 cm-1), G band (~1580 cm-1) and G' band (~2700). Good crystalline quality of pristine monolayer graphene can be deduced by a very weak disorder-induced D band, which is related to the non-sp2 bonding, and a relatively strong G band, which indicates sp2 symmetry. The signature of monolayer graphene can be seen in a single Lorentzian profile of the G' band with a sharp line width (~38 cm-1). The peak intensity ratios of I_D/I_G and $I_G/I_{G'}$ are measured to be 0.09 and 0.20 respectively.

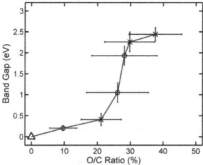

Figure 4. Correlation between band gap opening of oxidized graphene and the degree of oxidation (atomic ratio of oxygen to carbon atoms). Triangle, circle and asterisk marks indicate the pristine graphene, UV/ozone treated graphene and oxygen plasma treated graphene.

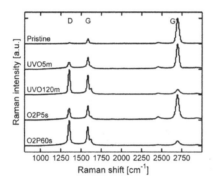

Figure 5. Typical Raman spectra of graphenes before and after oxidation process by UV/ozone treatment and oxygen plasma. All spectra are obtained with $E_{excit} = 2.41$ eV and normalized by the strongest peak.

As expected, a much stronger D band appears after the oxidation process takes place, where its peak intensity is varied by the oxidation level. Regardless of the oxidation process, a stronger D band peak is always found on sample that has higher oxygen/carbon atomic ratio. An increase of I_D/I_G ratio to 0.45 and 0.74 are observed for UVO5m and O2P5s samples respectively. For a prolonged oxidation process, a very high I_D/I_G ratio of 1.28 for both UVO120m and O2P60s samples is observed. The Raman spectrum of strongly oxidized graphene also shows a shoulder D' peak (~1620 cm-1) and decreases of G' band intensity. A significant increase of $I_G/I_{G'}$ ratios to 4.64 and 4.37 are observed for UVO120m and O2P60s samples respectively.

The increase of I_D/I_G ratio can be correlated to the increase of oxygenated functional groups bonded to the carbon atoms, which indicates the increase of perturbation to the lattice symmetry

[17]. The increase of $I_G/I_{G'}$ ratio can be associated to the increase of strain on the lattice induced by external forces [15], which in this case is caused by the increase of perturbation by oxygenated functional groups. At higher oxidation level, the extra strain induced by this perturbation is strong enough to break the lattice symmetry. This finding suggests that this symmetry breaking phenomenon is responsible for a sudden jump in band gap opening at oxygen/carbon atomic ratio of 25-30% (figure 4).

CONCLUSIONS

Oxygen plasma and UV/Ozone treatments as alternative methods for preparing graphene oxide have been studied. STM imaging shows islands of lattice disorder on the oxidized graphene. As is shown by scanning tunneling spectroscopy study, prolonged exposure of graphene to oxidizing treatments eventually creates a band gap opening around the zero-bias energy level. This energy band gap infers transformation of electronic properties of graphene from metallic to semiconducting as graphene undergoes higher degree of oxidation. Dry oxidation processes such as oxygen plasma and UV/Ozone treatments, could serve as promising methods for fine tuning the electrical properties of graphene oxide. They have the potential to replace conventional wet oxidation methods which utilizes harsh chemicals and requires more laborious efforts and time.

REFERENCES

[1]   A. Geim, and K. Novoselov, Nature Materials **6**, 183 (2007).
[2]   A. C. Neto *et al.*, Reviews of modern physics **81**, 109 (2009).
[3]   J. Oostinga *et al.*, Nature Materials **7**, 151 (2008).
[4]   K. Nakada *et al.*, Physical review. B, Condensed matter **54**, 17954 (1996).
[5]   M. Han *et al.*, Physical Review Letters **98**, 206805 (2007).
[6]   L. Ponomarenko *et al.*, Science **320**, 356 (2008).
[7]   I. Gierz *et al.*, Nano letters **8**, 4603 (2008).
[8]   T. Wehling *et al.*, Nano letters **8**, 173 (2008).
[9]   Z. Luo *et al.*, Applied physics letters **94**, 111909 (2009).
[10]   W. Hummers, and R. Offeman, Journal of the American Chemical Society **80**, 1339 (1958).
[11]   S. Park, and R. Ruoff, Nature nanotechnology **4**, 217 (2009).
[12]   X. Sun *et al.*, Nano Research **1**, 203 (2008).
[13]   D. Li *et al.*, Nature nanotechnology **3**, 101 (2008).
[14]   H. Becerril *et al.*, ACS nano **2**, 463 (2008).
[15]   M. S. Dresselhaus *et al.*, Nano letters **10**, 751 (2010).
[16]   A. Reina *et al.*, Nano letters **9**, 30 (2008).
[17]   A. Nourbakhsh *et al.*, Nanotechnology **21**, 435203 (2010).

Poster Session: Characterizations and Properties of Low-Dimensional Nanocarbon Structures

Mater. Res. Soc. Symp. Proc. Vol. 1284 © 2011 Materials Research Society
DOI: 10.1557/opl.2011.649

Thermal, Chemical and Radiation Treatment Influence on Hydrogen Adsorption Capability in Single Wall Carbon Nanotubes

Michail Obolensky[2], Andrew Kravchenko[2], Vladimir Beletsky[2], Yuri Petrusenko[3], Valeriy Borysenko[3], Sergey Lavrynenko[3], Andrew Basteev[1] and Leonid Bazyma[1]

[1]National Aerospace University, Kharkov Aviation Institute "KhAI", 17 Chkalov St., Kharkov, 61070, Ukraine
[2]V.N. Karazin Kharkiv National University, 4 Svobody Sq., Kharkov, 61077, Ukraine
[3]National Science Center - Kharkov Institute of Physics and Technology, 1 Akademicheskaya St., Kharkov, 61108, Ukraine.

ABSTRACT

The raw single wall carbon nanotubes (SWCNT) were chemically and thermally treated and then milled in ball mill. After this SWCNT were irradiated by electron beam with energy 2.3 MeV up to fluence 10^{14} e⁻/cm² at room temperature. Then SWCNT were saturated with hydrogen at pressure 5 bar and quenching down to the temperature 78 K. The sorption capability was measured by means of mass-spectroscopy and volumetric methods. The double increasing of mass hydrogen content in electron bombarded SWCNT was showed comparatively with non-irradiated samples.

INTRODUCTION

The main methods of carbon nanotubes (CNT) synthesis give the formation of products mixture with different degree of defect, structure and dimensional properties. The results reproducibility remains the serious problem of CNT practical use. This problem is linked with the requirement to CNT to have the same or at least very clear properties after purification and separation procedure. Despite the fact that there are a lot of different technologies for CNT synthesis now none of the above technologies provide the results reproducibility. Thus the procedures of CNT purification and separation have the primary importance.

As is well known the CNT thermal treatment enhance the hydrogen sorption capability sufficiently even at atmospheric pressure [1]. The chemical processing without thermal treatment can both enhance the hydrogen content as well decrease it [2]. The complex CNT treatment (chemical and thermal) are allow obtaining more significant values of sorption capabilities. The four times increasing of hydrogen percentage in comparison with non-treated samples was showed in [2] and hydrogen mass content was performed at the level more than 6 percent.

Enough high values of hydrogen mass content at relatively low pressures were demonstrated in [2] and this fact could be linked with defects presence in nanotube wall structure. Formation of these dislocations could be explained by annealing procedure for example and they form the singular wall shape and surface roughness which are associated with heightened hydrogen atoms binding energy [2].

The recent investigations [3-5] showed that there is the perspective for CNT sorption capability enhancing due to especial dislocation formation in it structure by means of electron bombardment and γ – quantum irradiation. The types of different dislocations which are appearing after SWCNT electron bombardment were considered especially in [6-7]. The experimental investigations of above mentioned factors (chemical and thermal treatment) as well the result of irradiation on to SWCNT sorption capability are performed in this work.

EXPERIMENT

Two technologies for CNT purification and separation were used. In accordance with the first method the SWNT (1 g amount, Carbon Solutions) was annealed in air at the temperature 350 °C during 30 minutes and after this was heated at the temperature 130 °C in 2.6 mole HNO_3 during 28 hours. Under the second method the 0.5 ml of 10 percent solution of $H[AuCl_4]$ was added to the annealed in air CNT with the aim to intensify the selectivity of amorphous carbon elimination. Then the 100 ml of concentrated $H_2SO_4 + H_2O_2$ (7:3) was added and the mixture was heated up to the temperature 90 °C with simultaneous interfusion. CNT processed by above procedure were flushed by methanol and then by water to achieve the neutral reaction, were dried in air and annealed in deep vacuum (10^{-8} bar) during 20 hours at temperature 1000 °C. The results of further control are evidencing about the second method advantages.

After the chemical treatment CNT were milled in stainless steel cryogenic "ball mill" equipped by outward thermo-insulation at liquid nitrogen temperature. Milling duration was 60 minutes with interruption after each 5 minutes with the aim to CNT samples microscopic control. After the milling procedure finishing the "ball mill" was heated up to the room temperature and CNT samples were bolted through the 63 μm sieve.

The method for SWNT electron bombardment and γ – quantum irradiation was elaborated. In contrast to [5] where the exposure of γ – quantum 10^5 Rad was used (this dose is equal to regime of electron bombardment ~$3*10^{12}$ e-/см^2) the fluence in presented work was 10^{14} e⁻/cm^2. The electron bombardment procedure was realized at room temperature on linear accelerator ELIAS (National Science Center "Kharkov Institute of Physics and Technology"). The electron energy was 2.3 MeV and beam intensity was 0.2 μA /cm^2.

Studying of sorption/desorption process has been fulfilled with use of standard volumetric method on the designed and manufactured vacuum stainless steel facility which included the known good unit volume V_E and measuring vessel with volume V_M. Part of measuring vessel V_{78} was cooled down to the temperature of liquid nitrogen boiling (~ 78 K). Before measurements the facility calibration was done. The good unit with volume V_E was inflated by gaseous helium or hydrogen with pressure P_I ~ 10 bar at room temperature which has been measured with precision ± 0.5 K. The pressure was controlled by electronic manometer GE Druck 104 with resolution 1 mbar. After total facility volume fill up $V = V_E + V_M$ the pressure dropped to value $P_F = V_E P_I (V_E + V_M - V_A)$. This circumstance allows determine the volumes relation exactly taking into account the ampoule volume V_A without sample. Above relation for hydrogen and helium was found equal with accuracy approximately ~ 0.01 percent and the hydrogen sorption by vacuum system elements could be excluded from consideration. After this procedure the container with ampoule was cooled to temperature 78 K and relation P_W/P_F was determined, where P_W is the pressure in the system after cooling.

The measurement of physically absorbed hydrogen was conducted in accordance with following procedure. First of all after container cooling down to temperature 78 K the system was evacuated to pressure ~ 0.1 mbar for gaseous hydrogen elimination. The next step was pressure increasing registration in system during container heating and caused by hydrogen desorption process. Hydrogen desorption out from SWNT was also studied by mass-spectrometry method on the mass spectrometer MX 7203.

DISCUSSION

The samples of SWNT with mass 80 – 300 mg (estimated bulk density is ~ 1000 kg/m^3) were used for experiments on the volumetric facility. The pressure drop character dynamics during container with non-irradiated samples cooling is performed in figure 1.

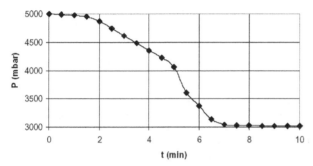

Figure 1. The dependence of pressure drop in the container with non-irradiated CNT samples upon time during container with samples cooling (P_F = 5035 mbar).

On the figure one may see three sections with different sorption dynamics. On the base of comparison of this pressure drop dynamics (see figure 1) with pressure drop dynamics for calibration stroke we can note following. The pressure drop relatively the empty container was observed already at room temperature during the letting-to-hydrogen and this pressure was installed at level ~ 20 mbar. The additional pressure decreasing was observed during further container cooling to temperature 78 K and at different cooling cycles this value was 10 – 15 mbar.

The pressure in the system during whole system heating was less than the pressure before cooling beginning on 10 – 12 mbar. Apparently this fact could be explained by any amount of hydrogen staying in binding state at room temperature after desorption. We didn't register this phenomenon at repeating cycles (cooling-heating).

Figure 2 illustrates the correlation of pressure and temperature in system for two evacuating regimes: fast (3 min) and gradual (15 min). Apparently already at temperature 78 K the increasing of evacuation duration causes the significant desorption and consequently the elimination of part of hydrogen out from system yet before heating beginning. The duration of container heating on both regimes was approximately equivalent 25 – 30 min.

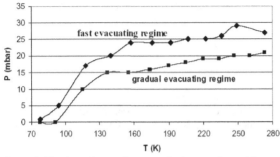

Figure 2. The dependence of pressure in the container with non-irradiated CNT samples upon temperature during container with samples heating for different evacuation rate.

The temperature dependences of hydrogen density in volume situated at heating regime (V_{78}) and in the rest facility volume at environment temperature are performed on figure 3. It could be seen that the main hydrogen mass is exuded at temperature T< 160 K that is caused by relatively low value of physical absorption activation energy. Pressure difference occurring at container heating without evacuating and initial pressure P_F as has been noted lately is caused by other mechanism with higher values of desorption energies and it can be related to chemosorption regime with higher character temperatures.

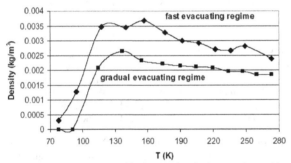

Figure 3. The dependence of hydrogen density in container with non-irradiated CNT samples upon temperature during container with samples heating for different evacuation rate.

Our estimations show that at different sorption/desorption cycles the amount of hydrogen located in non-irradiated SWNT was equal to 0.12 ±0.2 mass percent. Hydrogen desorption from treated and exposed to physical sorption/desorption procedure material at pressures ~3000 mbar as it is described above has been studied with use of mass-spectrometry method within the temperature interval 0 – 900 °C on the mass spectrometer MX 7203. The dependencies of amount of hydrogen extracted from non-irradiated (a) and irradiated up to fluence 10^{14} e⁻/cm² (b) samples upon temperature are showed in figure 4 in arbitrary units.

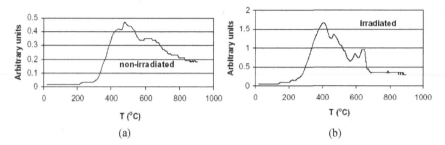

(a)   (b)

Figure 4. The dependencies of amount of hydrogen (arbitrary units) exuded from non-irradiated (a) and irradiated up to fluence 10^{14} e⁻/cm² (b) samples upon temperature.

The analogue spectrum for other extracted components with different mass numbers is performed on figure 5. The total amount of hydrogen relatively other desorbed gases are negligible (not more then 1 percent) so the performed spectrum is valid for both irradiated and non irradiated samples.

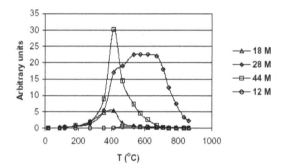

Figure 5. The dependencies of amount of gaseous components (arbitrary units) with different mass numbers exuded from CNT upon temperature.

As can be seen from figure 4 the amount of hydrogen desorbed from irradiated material is in 2.5 times more than such amount desorbed from non-irradiated. We have to draw attention on the fact that all procedures concerned to the sorption/desorption processes was executed at temperature interval 78 – 300 K but in the same time the significant amount of hydrogen is desorbed at temperatures more than 300 K. There are additional peaks on the desorption curve which apparently corresponds with different sorption sites appearing as the result of irradiation and can be characterized by various activation energies. It should be noted that over a long period of time of samples staying in air between irradiation procedures and hydrogen saturation the complete generated by irradiation sites filling by molecules of other gases has not been detected.

CONCLUSIONS

The amount of hydrogen desorbed from non-irradiated samples was 0.12 ±0.2 mass percent. The significant increasing (more than in two times) of hydrogen mass absorbed in irradiated SWCNT samples relatively non-irradiated ones has been proofed.

ACKNOWLEDGMENTS

Given research was encouraged by Science and Technological Center in Ukraine, project #4957.

REFERENCES

1. A. Anson, M.A. Callejas, A.M. Benito, W.K. Maser, M.T. Izquierdo, B. Rubio, J. Jagiello, M. Thommes, J.B. Parra and M.T. Martinez, *Carbon* **42**, 1243 (2004)
2. B.K. Pradhan, A.R. Harutyunyan, D. Stojkovic, J.C. Grossman, P. Zhang, M.W. Cole, V. Crespi, H. Goto, J. Fujiwara and P.C. Eklund, *J. Mater. Res.* **17**, 2209 (2002).
3. L.G. Zhou, F.Y. Meng and S.Q. Shi, *Nanotechnology* **3**, 154 (2003).
4. Wen Qian, Jiafu Chen, Lingzhi Wei, Liusuo Wu and Qianwang Chen, *NANO* **4**, 7 (2009).
5. M.A. Obolensky, A.V. Basteev and L.A. Bazyma, *Fullerenes, Nanotubes, and Carbon Nanostructures*, **19**, (2010) (in press)
6. F. Banhart, *Fiz.Tv. Tela.* **44**, 388 (2002)
7. A.V. Krasheninnikov and K. Nordlund, *Fiz. Tv. Tela.* **44**, 452 (2002)

Mater. Res. Soc. Symp. Proc. Vol. 1284 © 2011 Materials Research Society
DOI: 10.1557/opl.2011.650

Electrogenerated chemiluminescence from carbon dots

L. Sun[1,2], T.-H. Teng[1], Md. H. Rashid[1], M. Krysmann[1], P. Dallas[1], Y. Wang[1], B.-R. Hyun[2], A. C. Bartnik[2], G. G. Malliaras[1], F. W. Wise[2], E. P. Giannelis[1]
[1]Material Science and Engineering Department, [2]School of Applied and Engineering Physics, Cornell University, Ithaca, NY14853, USA

ABSTRACT

We report an interesting property of carbon dots: they emit light under charge injection. We synthesized carbon dots in diameter about 20 nm using wet chemistry methods. The photoluminescence quantum efficiency of the carbon dots dissolved in water was about 11%. We observed strong electrogenerated chemiluminescence (ECL) from the sample. This observation of ECL from carbon dots indicates that they could be a good candidate material for carbon-based electroluminescent devices.

INTRODUCTION

Fluorescent semiconductor quantum dots have generated broad promising applications including biological labeling and solid-state lighting. Nanometer-size carbon dots which are the counterparts of silicon nanoparticles now light up. Although the typical photoluminescence quantum efficiency of carbon dots is not high yet, they are non-blinking [1] and have large two-photon absorption cross section [2] which are favorable for one-photon or two-photon bio-imaging. On the other hands, getting luminescence through charge injection has been tried by several groups [3], which is a step towards electroluminescence from carbon dots. Carbon dots could be a good candidate material for nontoxic carbon-based optoelectronic devices.

There are two schools of fabrication methods for carbon dots: top-down, bottom-up. Top-down method starts from bulk carbon, and using laser [2] or electrochemical oxidation [4] to break down the carbon into small pieces which have the size in nanometer scale. Graphene fragment carbon dots exhibit size tunable optical properties [5]. The bottom-up method starts from organic molecules and using chemical reaction, pyrolysis method to fabricate carbon dots, pioneered by Giannelis groups [6]. Following this direction, we can make mass production of carbon dots. Above on this, we investigate their optical properties and observed bright electrogenerated-chemiluminescence (ECL) from carbon dots dispersed in acetonitrile solution. Moreover, ECL from carbon dot film sticking to the surface of mesoporous TiO_2 is also observed. The spectrum of the ECL peaks at about 460 nm. The results indicate that the carbon dots we synthesized could be a good material for blue emitting LEDs.

EXPERIMENT

Carbon dot synthesis

The method of synthesis of C dots is adopted from a previously reported method with some modification [6]. In a typical synthesis; a mixture of citric acid or acetylenedicarboxylic acid were mixed with ethanolamine under constant magnetic stirring to get a mole ratio of amine to acid of 3: 1 or 2: 1. After homogeneous mixing the reaction mixture was heated to above 70 °C to get a syrupy suspension which is pyrolyzed directly in a furnace at temperature more than 180 °C. The black product is highly water soluble and is used for different spectroscopic and microscopic analysis.

Electron microscopy

After dialysis, the unreacted precursors were filtered out. The carbon dots exhibit quite uniform size under electron microscope (as shown in Figure 1). No lattice structure was observed from the carbon dots. They are basically amorphous.

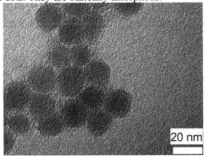

20 nm

Figure 1. TEM image of carbon dots. Their size is about 20 nm.

Optical absorption and emission

The carbon dots do not exhibit many optical absorption peaks as small molecules, rather they show continuous increasing of the absorbance while the wavelength decreases (shown in Figure 2). A weak absorption peak at around 350 nm is observed, which might tell the intrinsic energy gap in the material. The PL spectrum has a peak at around 460 nm while the dots are excited with light source of wavelength less than 400 nm. The PL peak position does not change much when the wavelength of the exciting light source is less than 400 nm, but red-shifts when the exciting wavelength is longer than 400 nm. As a control, PL spectrum from the solution without carbon dots was measured, which showed no observable signal in the same condition.

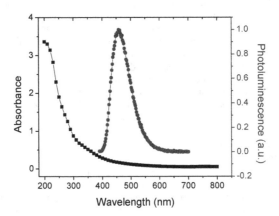

Figure 2 Absorption (solid squares) and emission (solid circles) spectra of carbon dots dispersed in water.

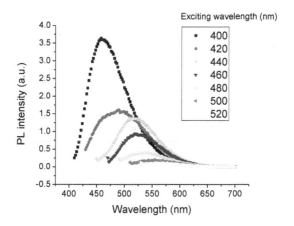

Figure 3 Photoluminescence from carbon dots dispersed in water, excited at different wavelength.

The absolute photoluminescence quantum yield of the carbon dots dispersed in water was measured using an integrating sphere [7], which is based on the design proposed by Friend et al. [8] The typical quantum yield is about 11% when the carbon dots are excited by light source at 400 nm wavelength.

Electrogenerated chemiluminescence

The experiment setup for measuring the luminescence through charge injection was similar to a typical cyclic voltammetry measurement [9]. The sample was prepared by dissolving carbon dots into an acetonitrile solution with tetra-n-butylammonium perchlorate (TBAP) as the electrolyte. A glassy carbon disk was used as the working electrode while a platinum wire and a silver wire served as the counter and the quasi-reference electrodes, respectively. A potentiostat (AFRDE5, Pine Instrument Company) was employed to supply and measure the electric potential and current. A photomultiplier tube (PMT, Hamamatsu R442) operated in either current or photon-counting mode was mounted facing the working electrode to detect the ECL photons emitted by the carbon dots. The cyclic potential signal and the corresponding current signal were digitized and recorded by a computer, which also simultaneously recorded the number of photons detected by the PMT. All ECL measurements were carried out under ambient conditions.

The potential on the working electrode was scanned between +2.8 V and -2.8 V relative to Ag/AgCl. As shown in Figure 4a, the ECL signal appeared when the current increased sharply, indicating the relation between light emission and charge injection. The ECL signal is quite stable during the electrochemical process in about 10 minutes. Besides the ECL peaks at the maximum (+2.8 V) or minimum potential (-2.8 V), two other ECL peaks appeared at potentials about -1.5 V and +2.1 V respectively (as shown in Figure 4b). The difference of the two potentials is about 3.6 V, which is consistent with the energy gap of the carbon dots derived from the optical absorption data.

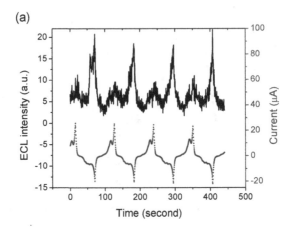

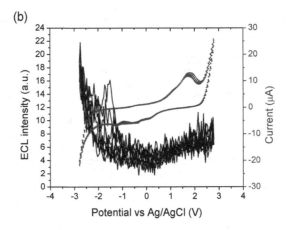

Figure 4 (a) Time dependent ECL intensity (solid line) and current (dots) from carbon dots when the potential scanned from -2.8 V to +2.8 V relative to Ag/AgCl reference electrode. (b) ECL intensity (solid line) and current (dots) dependence on the potential.

To further investigate the nature of the ECL from the carbon dots, we measure the spectrum of ECL light. As shown in Figure 5, the ELC spectrum has a peak at 460 nm, overlapping with the PL spectrum (excited at 400 nm) very well. This demonstrates that the exciton formed through charge injection is basically the same as through optical pumping. At longer wavelength, we also observed non-vanishing ECL signal, while PL signal almost vanished. Since ECL is more sensitive to surface chemistry [9, 10, 11, 12, 13], it is not surprising that it can probe more surface states than PL. We didn't observe significant red shift of ECL peak relative to PL peak (excited at 400 nm or lower), which is very different from Si nanocrystals [10]. This result indicates that there are much less emissive surface states on the carbon dots.

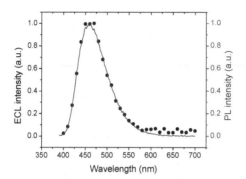

Figure 5. ECL and PL spectra of carbon dots dispersed in acetonitrile containing 0.1 M TBAP.

A bright ECL from film state carbon dots was also observed in our experiment. We prepared the sample by adsorbing carbon dots onto mesoporous TiO_2 film. A bare mesoporous TiO_2 film was used as a control. Though TiO_2 film emit light itself under cyclic potential between -2.5 V to +2.5 V, the TiO_2 film covered by carbon dot film showed much stronger ECL signal (shown in Figure 6). The results indicate that majority of the ECL signal is from the carbon dot film. Further spectroscopic study of the ECL signal can tell them apart in a rigorously way. The strong ECL from the carbon dot film tells that it is promising to make electroluminescent thin film devices using carbon dots.

DISCUSSION

Though the progress made in carbon dots synthesis and characterization, the mechanism of light emission from carbon dots is still under debating. Though it is possible the photoluminescence is from the core of the carbon dot where the optical absorption mainly takes place, other evidences also show the possibility of surface light emission. In this situation, ECL which is sensitive to surface property of nanoparticles could sever as a good tool to clarify the light emission mechanism. The ECL from carbon dots could also be used for immunoassay, due to their intrinsic non-toxicity. It is also promising to make carbon dots based optoelectronic devices, since we have successfully demonstration ECL from carbon-dots film.

CONCLUSIONS

We have observed ECL from carbon dots either dispersed in organic solvent or in a film state. The spectrum of ECL shows strong peak overlapping with PL peak. ECL signal at longer wavelength was also observed which was attributed to surface states of the carbon dots.

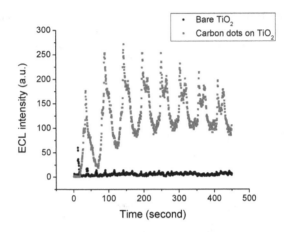

Figure 6. ECL from bare mesoporous TiO₂ (solid circles) and carbon dots covered mesoporous TiO₂ (solid squares) when the potential was scanned from -2.5 V to +2.5 V relative to Ag/AgCl reference electrode. The sample was immersed in acetonitrile containing 0.1 M TBAP.

ACKNOWLEDGMENTS

We acknowledge support from Award No. KUS-C1-018-02, made by King Abdullah University of Science and Technology (KAUST). We also acknowledge support by the National Science Foundation under Grant No. EEC-0646547, NYSTAR and CCMR at Cornell University.

REFERENCES

1. Y.-P. Sun et al, *J. Am. Chem. Soc.* 2006, 128, 7756
2. L. Cao et al, *J. Am. Chem. Soc.* 2007, 129, 11318
3. L. Zheng et al, *J. Am. Chem. Soc.* 2009, 131, 4564
4. J. Lu et al, *ACS Nano* 2009, 3, 2367
5. H. Li et al, *Angew. Chem. Int. Ed.* 2010, 49, 4430
6. A. B. Bourlinos et al, *Small* 2008, 4, 455
7. L. Sun et al, *Small* 2010, 6, 638
8. J. C. De Mello, H. F. Wittmann, R. H. Friend, *Adv. Mater.* 1997, 9, 230
9. L. Sun et al, *Nano Lett.* 2009, 9, 789
10. Z. Ding et al, *Science* 2002, 296, 1293
11. N. Myung et al, *Nano Lett.* 2004, 4, 183
12. N. Myung et al, *Nano Lett.* 2002, 2, 1315
13. N. Myung et al, *Nano Lett.* 2003, 3, 1053

Mater. Res. Soc. Symp. Proc. Vol. 1284 © 2011 Materials Research Society
DOI: 10.1557/opl.2011.365

Far Infrared Characterization of Single and Double Walled Carbon Nanotubes

S. G. Chou[1], Z. Ahmed[1], G.G. Samsonidze[2], J. Kong[3], M. S. Dresselhaus[3, 4], D. F. Plusquellic[1]
1 Physical Measurement Laboratory, National Institute of Standard and Technology, Gaithersburg, MD20899
2. Department of Physics, University of California and Materials Sciences Division, Lawrence Berkeley National Laboratory, Berkeley, CA94720
3. Department of EECS and 4. Department of Physics, Massachusetts Institute of Technology, Cambridge, MA02139

ABSTRACT

High resolution far infrared absorption measurements were carried out for single walled and double walled carbon nanotubes samples (SWCNT and DWCNT) encased in a polyethylene matrix to investigate the temperature and bundling effects on the low frequency phonons associated with the low frequency circumferential vibrations. At a temperature where $k_B T$ is significantly lower than the phonon energy, the broad absorption features as observed at room temperature become well resolved phonon transitions. For a DWCNT sample whose inner tubes have a similar diameter distribution as the SWCNT sample studied, a series of sharp features were observed at room temperature at similar positions as for the SWCNT samples studied. The narrow linewidth is attributed to the fact that the inner tubes are isolated from the polyethylene matrix and the weak inter-tubule interactions. More systematic studies will be required to better understand the effects of inhomogeneous broadening and thermal-excitation on the detailed position and lineshape of the low frequency phonon features in carbon nanotubes.

INTRODUCTION

The detailed phonon structure of carbon nanotubes determines the mechanical and thermal transport properties for nanotube-based materials and devices[1]. At room temperature, the thermal transport properties of nanotubes are mostly determined by optical phonons, whereas at a temperature significantly lower than the Debye temperature, the detailed thermal properties of carbon nanotubes are largely determined by low-frequency, acoustic-branch phonons[2, 3]. In most vibrational spectroscopic studies of SWCNT, the energies for phonons could be derived from the zone folding scheme, whereby the phonon dispersion relation of 2D graphene is folded into the 1D Brillouin zone for the SWCNTs[4]. The approach, in most cases, yields predictions that are consistent with both resonance Raman and IR spectroscopic findings [5, 6]. On the other hand, special theoretical considerations are required in the case of the lower frequency modes derived from the acoustic branches of graphene because some of the in and out of plane acoustic branch modes associated with graphitic translation in nanotubes cannot be expressed exactly in the zone folding scheme[4]. A number of theoretical models have been developed based on elasticity theory and mass-spring constants to predict the structures of these phonon modes in the low energy region for single walled carbon nanotubes, usually associated with circumferential vibrations, but the detailed experimental measurements of the individual phonon transitions in the far infrared (FIR) energy region remain challenging[3, 7, 8]. Since the phonon energies associated with the acoustic branch phonons are small, the absorption cross sections resulting from the electric dipoles generated by vibrating carbon atoms associated with these modes are small. As the carbon atoms interact with their environments through bundling or substrate interactions, the low frequency phonon transitions become further damped. Also, at room temperature, with the thermal energy (25meV) much larger than the acoustic-branch phonon energies, the individual transitions usually could not be resolved due to the thermally excited phonon population. In this report, we use a high resolution continuous wave THz setup to characterize the low energy transitions of carbon nanotubes in the FIR region. At a temperature where k_BT is significantly lower than the phonon energy, most of the thermally excited phonon population relaxes down to the ground states, and the broad absorption features as observed at room temperature for SWCNTs become well resolved phonon transitions. To further understand the effects inhomogeneous broadening of these low energy phonons due to bundling and surface interactions, we study a DWCNT sample whose inner tubes have a similar diameter distribution as for the SWCNT studied. A series of very sharp phonon features were observed at room temperature at similar positions as for the SWCNT samples. The narrow linewidth is attributed to the isolated environment the inner tubes are situated in, but more systematic, temperature-dependent studies on both the DWCNT and SWCNT samples with better-defined diameter distributions will be required to understand the detailed phonon interactions in these samples.

EXPERIMENTAL

A detailed description of the continuous wave THz spectroscopy setup has been described in detail elsewhere[9]. In the present work, the THz excitation was generated by a tunable Ti:sapphire laser, where frequency selection is achieved by coupling through an end mirror to an external grating-tuned cavity. The 14 cm wide, 1800 line/mm gold-coated grating is configured to scan ~100 cm^{-1} at a wavenumber resolution of 0.03 cm^{-1}. The spectrometer employs an ErAs\GaAs photomixer. Output from a fixed frequency Ti:sapphire ring laser near 11800 cm^{-1} is combined with the output from the grating tunable Ti:sapphire laser and focused onto the photomixer, driving the antenna structure to emit the difference frequency which sweeps through 3 THz (100 cm^{-1}) as the grating tunes the Ti:sapphire laser. The THz radiation is collimated by a Si lens, and then focused by off-axis parabolic mirrors. The sample assembly, fixed to the cold finger of a liquid He cryostat in the vacuum chamber initially maintained at (6-7) pascal prior to cooling, is placed in the THz beam, and the transmitted light is detected by a liquid He cooled silicon composite bolometer. For normalization purposes, a voltage proportional to the photomixer photocurrent is recorded simultaneously with the detected THz signal. The carbon nanotube samples consist of pellets with several percent of carbon nanotubes in 100 mg of polyethylene (PE) matrix. The relative concentration of the nanotubes in the PE matrix is adjusted to maximize the signal to noise ratio. The CG100 CoMoCAT SWCNT sample used in the experiment was purchased from Southwest Technologies. The DWCNT sample was manufactured by Helix Materials Solutions. Both samples were used as-is without further purification. The FIR absorption spectra of SWCNT-containing pellets are normalized to a blank PE pellet. The Raman characterization was performed in the back-scattering geometry using the 676 nm excitation from a Kr laser through a long distance 100x objective. The laser excitation power was kept below 0.5 mW/μm² on the sample at all times to avoid excessive heating.

RESULTS AND DISCUSSION

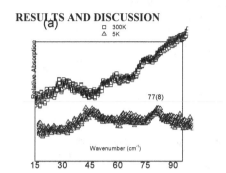

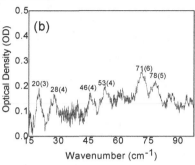

Figure 1. (a) Far IR absorption of CoMoCAT SWCNTs at room temperature (red) and 4K (blue) ((b) Far IR Absorption of DWCNTs measured at room temperature. The spectral profiles were fitted to Gaussian lineshapes. The center of the peak for the individual transition is labeled in the spectra, whereas the width of the peaks are included in the parentheses.

Figure 1(a) shows the FIR absorption spectra of the SWCNT samples at room temperature and at 5 K. At room temperature, the absorption spectrum for the SWCNTs is consistent with previous time-domain THz absorption studies of SWCNTs with a broad and rising baseline and non-distinct transitions seen at 35 cm⁻¹ and 76 cm⁻¹. At 5 K, by reducing the thermally excited phonon populations, these broad transitions are resolved into clearly-defined vibrational features with linewidths of around 10 cm⁻¹. The lineshape analysis of the peaks shows Gaussian-type profiles, suggesting that the features are still significantly broadened, most likely by bundling and by the surface interactions between the nanotube and the PE matrix. As the temperature drop to 5K, a blue shift from 35 to 45 cm⁻¹ is also observed for the lower frequency mode. The shift is consistent with the condensation of O_2 gas adsorbing on SWCNTs at low temperature. At an initially low vacuum condition of (6-7) pascal, where our experiment is carried out, previous nanotube photoluminescence studies have suggested that a significant amount of gas adsorption is present at room temperature[10], which could change the reduced mass on nanotube surface and shift the lower frequency modes.

To further investigate the low frequency features and the effects of surface interactions, a DWCNT sample whose inner tubes have a similar diameter distribution to the CoMoCAT SWCNT sample was studied. Since the energies of the low frequency phonons are more dependent on diameter rather than on chirality, the inner-walled nanotubes of DWCNT sample with a similar diameter distribution could provide insights into the vibrational transitions without the artifacts associated with bundling and nanotube-matrix interactions. Figure 2 shows the Raman spectra for the radial breathing mode (RBM) region of the two samples, excited at 676 nm. Since the diameter of the SWCNTs is inversely proportional to the RBM frequency, the positions of the RBM could give a general idea of the diameter distribution of the two samples. Comparing the group of RBM frequencies centered around 290 cm^{-1} for the two samples, the inner-most layer for the DWCNT sample appears to have a similar diameter distribution to the SWCNT sample, which is around 0.85nm (while the outer layer(s) may vary).

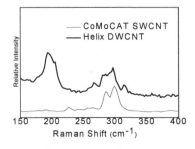

Figure 2. Spectra for the radial breathing mode (RBM) region of the two samples, excited at 676nm.

Figure 1(b) shows the absorption spectrum of the DWCNT sample measured at room temperature. At room temperature, a series of sharp peaks emerged with a linewidths of 3 cm^{-1} to 6 cm^{-1}, which is comparable to the linewidth of Raman features of individual nanotubes isolated on a SiO_2 substrate. The narrow linewidth suggests that the peaks arise from the inner tubes of the DWCNT sample, which are situated in a more isolated environment. Since these inner tubes have a similar diameter distribution to the SWCNT sample, we expect the position of the peaks shown in Fig. 1 (b) to be similar to the ones shown in Fig. 1 (a). Six stronger transitions were observed for DWCNT sample at 20 cm^{-1}, 28 cm^{-1}, 46 cm^{-1}, 53 cm^{-1}, 71 cm^{-1}, and 78 cm^{-1} while a number of weaker transitions were also present. Some of the main groups of peaks do appeared to have similar positions as the ones observed in the SWCNT sample, but the peaks observed in the DWCNT samples appear in groups of two, (7-8) cm^{-1} apart.

The group RBM peaks between 250 cm^{-1} and 320 cm^{-1} in Figure 2 suggests that the inner tubes of the DWCNT sample do have a wider diameter distribution compared to the CoMoCAT SWCNT sample. As a result, the dual-peak features as shown in Figure1 (b) could simply be a result of different nanotubes in the sample with similar diameters. It is also possible that the dual peak feature could be a result of the inter-layer interactions between the inner and outer tubes. A lineshape analysis shows that the transitions have Gaussian-type profile, suggesting that inhomogeneous broadening and thermally excited phonon populations still play an important role in the optical process. However, more systematic and detailed temperature-dependent studies on both SWCNT and DWCNT samples with more homogeneous diameter distributions will be required to better understand the pertinent interactions.

SUMMARY

In summary, high resolution far infrared characterization measurements were carried out for single walled and double walled carbon nanotubes samples (SWCNT and DWCNT) encased in a polyethylene matrix to investigate the spectral response to temperature and bundling effects with particular emphasis given to the low frequency phonons associated with circumferential vibrations. For SWCNTs, intertubule bundling and surface interaction effects were found to significantly broaden and dampen the optical absorption feature as much as the thermally excited phonon population effects. At a temperature where k_BT is significantly lower than the phonon energy, the broad absorption features as observed at room temperature become well resolved phonon transitions. On the other hand, for a DWCNT sample whose inner tubes have similar diameter distribution as the SWCNT sample, a series of sharp groups of features were observed at room temperature at similar positions as the SWCNT samples. The narrow linewidth is attributed to the isolated environment the inner tubes are situated in, but more systematic, temperature-dependent studies on both the DWCNT and SWCNT samples with better-defined diameter distributions will be required to understand the detailed phonon interactions in these samples.

ACKNOWLEDGEMENTS:

S. G. C. and Z. A. gratefully acknowledge support from the National Research Council Fellowship. S. G. C. gratefully acknowledges helpful discussion with Prof E. B. Barros and Prof. S. B. Cronin. The authors also gratefully acknowledge Dr. J. Fagan for providing us with the commercial grade Helix DWCNT samples. Certain equipment or materials are identified in this paper in order to specify the experimental procedure adequately. Such identification is not intended to imply endorsement by the National Institute of Standards and Technology, nor is it intended to imply that the materials or equipment identified are necessarily the best available.

REFERENCES:

1.   M. S. Dresselhaus and P. C. Eklund, Adv. Phys. **49** (6), 705-814 (2000).
2.   J. Hone, in *Carbon Nanotubes* (Springer-Verlag Berlin, Berlin, 2001), Vol. 80, pp. 273-286.
3.   J. Zimmermann, P. Pavone and G. Cuniberti, Phys. Rev. B **78** (4), 13 (2008).
4.   R. Saito, *Physical Properties of Carbon Nanotubes*, 1st edition ed. (Imperial College Press, London, 1998).
5.   U. J. Kim, X. M. Liu, C. A. Furtado, G. Chen, R. Saito, J. Jiang, M. S. Dresselhaus and P. C. Eklund, Phys. Rev. Lett. **95** (15), 4 (2005).
6.   G. G. Samsonidze, R. Saito, N. Kobayashi, A. Gruneis, J. Jiang, A. Jorio, S. G. Chou, G. Dresselhaus and M. S. Dresselhaus, Appl. Phys. Lett. **85** (23), 5703-5705 (2004).
7.   D. Gunlycke, H. M. Lawler and C. T. White, Phys. Rev. B **77** (1), 9 (2008).
8.   H. Suzuura and T. Ando, Phys. Rev. B **65** (23) (2002).
9.   K. Siegrist, C. R. Bucher, I. Mandelbaum, A. R. H. Walker, R. Balu, S. K. Gregurick and D. F. Plusquellic, Journal of the American Chemical Society **128** (17), 5764-5775 (2006).
10.   J. Lefebvre, P. Finnie and Y. Homma, Phys. Rev. B **70** (4) (2004).

Mater. Res. Soc. Symp. Proc. Vol. 1284 © 2011 Materials Research Society
DOI: 10.1557/opl.2011.223

Curvature-induced Symmetry Lowering and Anomalous Dispersion of Phonons in Single-Walled Carbon Nanotubes

Jason Reppert[1], Ramakrishna Podila[1], Nan Li[2], Codruta Z. Loebick[2], Steven J. Stuart[3], Lisa D. Pfefferle[2] and Apparao M. Rao[1]

[1]Department of Physics and Astronomy; Center for Optical Material Science and Engineering Technologies, Clemson University, Clemson, South Carolina, USA.
[2]Department of Chemical Engineering, Yale University, New Haven, Connecticut, USA
[3]Department of Chemistry, Clemson University, Clemson, South Carolina, USA

ABSTRACT

Here we report rich and new resonant Raman spectral features for several sub-nanometer diameter single wall carbon nanotubes (sub-nm SWNTs) samples grown using chemical vapor deposition technique operating at different temperatures. We find that the high curvature in sub-nm SWNTs leads to (i) an unusual S-like dispersion of the G-band frequency due to perturbations caused by the strong electron-phonon coupling, and (ii) an activation of diameter-selective intermediate frequency modes that are as intense as the radial breathing modes (RBMs). Furthermore, an analytical approach which includes the effects of curvature into the overlap integral and the energy gap between the van Hove singularities is discussed. Lastly, we show that the phonon spectra for sub-nm SWNTs obtained from the molecular dynamic simulations which employs a curvature-dependent force field concur with our experimental observations.

INTRODUCTION

Resonant Raman scattering of single walled carbon nanotubes (SWNTs) has been studied extensively [1]. When laser excitation energy (E_{laser}) matches the energy gap (E_{ii}) between a pair of ith van Hove singularities (vHS), one observes a resonance enhancement in the intensity of Raman-active modes such as the radial breathing mode (RBM) and the tangential stretching mode (G-band). E_{ii} is often obtained from the zone-folding scheme using cutting lines to deduce the 1D electronic energy sub-bands for SWNTs [1,2]. Within the framework of the zone-folding scheme, most Raman features of SWNTs whose average diameter is > 1 nm have been well understood [1,2].

Here, we present strong spectroscopic evidence which suggests that curvature-induced effects lead to the activation of resonant intermediate frequency modes (IFMs). Previously, extremely *weak* IFMs (600-1100 cm^{-1}) have been reported by several groups [3, 4] in SWNTs with average diameter in the range of 1.5±0.3 and 1.0±0.3 nm. For example, Fantini *et al.* observed a step-like dispersion of the IFMs in semiconducting and metallic SWNTs and attributed its origin to the creation of an optic phonon and annihilation of an acoustic phonon. We find that the intensity of the IFMs can equal that of the RBMs in chemical vapor deposition (CVD) grown sub-nm SWNTs. Interestingly, we also observe S-like dispersion for the G-band frequency which is a result of curvature-induced perturbation of the electron-phonon interaction. We show that by taking into account the chiral dependence of $E_{ii,}$ (obtained from the tight-binding method [4]) and the overlap integral γ_o, the origin of the observed Raman features in sub-nm SWNTs can be understood. Furthermore, from our molecular dynamic simulations which employ a curvature-dependent force field [5], we find that (i) the G-band dispersion consists of a family of lines with $2n + m$ = constant, where n and m are the standard chiral indices [1], and (ii) the intensity of the IFMs decreases with decreasing nanotube curvature.

EXPERIMENT

Nine different sub-nm SWNT bundles were synthesized using a thermal CVD process with either mono (Co) or bimetallic (CoMn) catalysts as described previously [6, 7]. Briefly, a CoMn-MCM-41 catalyst with 3% (or 1%) metal loading (Co:Mn in a molar ratio of 1:3 or 1:1) was synthesized by isomorphous substitution of metal in the silica framework. A 16 carbon atoms alkyl chain length was used to template the MCM-41 yielding an average pore diameter of about 3 nm as determined by nitrogen physisorption measurements. Sub-nm SWNT bundles were obtained from a thermal disproportionation of CO at 600 – 800 °C in a quartz tube reactor. The CoMn-MCM-41 catalyst (~200 mg) was reduced (flowing hydrogen, 700°C, 1 atm.) prior to the disproportionation of pure CO at 5-6 bar at desired synthesis temperature (see Fig. 1). Further details of synthesis and characterization of sub-SWNTs prepared using monometallic catalyst can be found in [8]. All samples used in this study exhibited similar rich spectral features. For simplicity, we focus on CVD-Co-Silica-550 which exhibited RBMs above 240 cm^{-1} confirming the presence of sub-nm SWNTs (average SWNT diameter ~ 0.75±0.1 nm) as shown in Fig. 2a. Sample ID CVD-Co-Silica-550 refers to the sub-SWNT bundles prepared at 550 °C using 3% Co catalyst supported on a grafted Silica wafer.

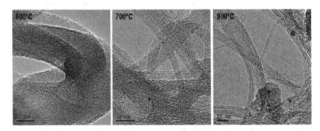

Figure 1: TEM of as-prepared sub-nm SWNTs synthesized at 600, 700, and 800 °C using CoMn-MCM-41 bimetallic catalyst. The scale bar shown in each micrograph is 10 nm.

DISCUSSION

In our earlier work using SWNT bundles prepared by the electric-arc and the pulsed laser vaporization methods the G-band frequency was found to be independent of the tube diameter. In contrast, the G-band frequency (both G$^-$ and G$^+$ modes) in Fig. 2c softens (downshifts) at first ($\lambda_{excitation}$ = 532 nm) and hardens (upshifts) with increasing excitation wavelength ($\lambda_{excitation}$ = 647 nm). With further increase in $\lambda_{excitation}$, it re-softens ($\lambda_{excitation}$ = 785 and 1064 nm) exhibiting an overall S-like dispersion. We relate this dispersion of the G-band frequency to electron-phonon interaction and perturbation of the optical phonon energy due to the *high curvature* in sub-nm SWNTs. In agreement with Sasaki *et al.* diameter dependence of the TO and LO phonon frequencies [9], we concur that the hardening (or softening) of the optical phonon frequencies

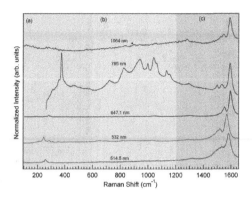

Figure 2: Raman spectra of CVD-Co-Silica-550 sub-nm SWNT samples. (a) Radial breathing modes at different excitation wavelengths. The 785 nm excitation is observed to be in resonance with 0.61 nm ($\omega_{RBM} \sim$ 374 cm^{-1}). (b) Intense intermediate frequency modes appear in the spectrum excited at 785 nm excitation. (c) S-like dispersion of the G-band frequency is observed as the excitation wavelength was varied from 514.5 to 1064 nm.

occurs due to nanotube curvature. Similar dispersions of the G-band have also been observed in other sub-nm SWNT samples used in this study. It is well known that the lineshape of the G-band deviates from a Lorentzian to a Breit-Wigner-Fano (BWF) lineshape in SWNT bundles due to charge transfer or the interaction of Raman-active phonons with the energy continuum near the Fermi energy, E_F [10, 11]. The BWF lineshape is described as a measure of the departure from a symmetric Lorentzian lineshape due to electron-phonon interaction [10]. Consistent with Sasaki *et al.* [9], sub-nm diameter tubes and highly curved bundles present in our samples can lead to strong electron-phonon coupling due to the broadening of the vHS [12], and an increase in the allowed phonon spectrum due to symmetry lowering [13]. Thus, the G-band in Fig. 2c obtained using various excitation wavelengths is fit to BWF lineshape [10] with $1/q$ as the only fitting parameter. As shown in Fig 3a, the obtained $1/q$ values clearly mimic the S-like dispersion of the G-band mentioned above. Furthermore, it is interesting to note that the observed frequency shifts in the RBM track the G-band frequency shifts, indicating that the observed G-band intensity is from different population of tubes with dissimilar internal strains and metallicity.

We further used molecular dynamics simulations on 30 different chiral and achiral tubes (Fig. 3b) with the well-known adaptive intermolecular reactive empirical bond order (AIREBO) potential [5] to confirm the dispersion of G-band due to high curvature. AIREBO is a reactive potential capable of treating variations in bond order and hybridization states. Although this semi-classical model contains no explicit treatment of electronic degrees of freedom, it does include the effects of SWCNT curvature on bond order and force constants. It is worth noting that the interaction is included via an empirical approach to the analytical bond- order theory, which is equivalent to orthogonal tight-binding theory in the second-moments approximation [14]. Thus, SWNTs with small radii have reduced carbon-carbon bond angles which lead to decreased bond orders and more sp^3-like bonding character. Importantly, such molecular dynamics simulations allow the effects of dynamic variation in local curvature to be included in the simulated power spectra.

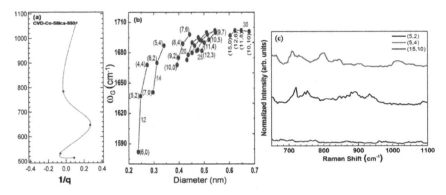

Figure 3: (a) Strength of electron-phonon interaction ($1/q$ values) obtained from BWF fit to the G-band (Fig. 2c) show S-like dispersion. (b) Dispersion of G-band obtained from molecular dynamics simulation that employs a curvature-dependent force field exhibits $2n + m$ family lines. (c) Intermediate frequency modes obtained from the molecular dynamic simulations for different chiralities indicate that the IFMs are relatively weak at lower curvatures as in (15, 10) tube compared to (5, 2) or (5, 4) tubes.

It is interesting to note that the dispersion of the G-band obtained from these simulations exhibits a $2n + m$ = constant family line behavior (Fig. 3b). As shown in Fig. 3b, the dispersion length of such family lines decreases as the value of $2n + m$ increases indicating that increase in the curvature can lead to a shift in the G-band as observed in Fig. 2c. In addition, as expected [15], the dispersion of the G-band is negligible for large diameter SWCNTs (for $2n + m$ = 30 as seen in Fig. 3b) and is maximum for sub-nm SWNTs ($2n + m$ = 12 family). Previously, Jorio *et al*, showed that the G$^+$ feature is both diameter and chiral angle independent in case of SWNTs with $d > 1$ nm. However, the G$^-$ features were found to be strongly diameter dependent [1]. As discussed above, in case of sub-nm SWNTs we observed that both these features exhibit dispersion with the incident laser energy due to high nanotube curvature (Fig. 2c).

Returning to Fig. 2b, we note that the IFMs of sub-nm SWCNTs are as intense as the RBMs. Fantini *et al*. [3] suggested that the IFMs result from a simultaneous creation of an optical phonon and annihilation (or creation) of an acoustic phonon. Importantly, a step-like dispersion of the IFMs was observed for tube diameters in the range ~1-1.5 nm [3]. Fantini *et al*. [3] observed a linear dispersion of the IFMs with a positive (negative) slope when an acoustic phonon was created (annihilated) simultaneously with an optical phonon. Thus, in case of [3], a plot of IFM frequencies as a function of the laser energy exhibits a V-shape (see Ref. 16). We find that in this study (i) several IFMs fall outside the proposed V-shape area, and (ii) the intensity of the IFMs are enhanced only for certain $\lambda_{excitation}$. In particular, intense IFMs are observed with the 785 nm excitation from 0.61 nm sub-nm SWNTs (CVD-Co-Silica-550; ω_{RBM} ~ 374 cm^{-1}) which are not resonant with any other excitation wavelengths used in this study.

In agreement with the Raman spectra, our molecular dynamic simulations that include the effects of curvature in the force field also show an enhanced intensity for the IFM modes in SWCNTs with higher curvature. For example, as shown in Fig. 3c, it is evident from the power spectrum of (15, 10) tubes that the IFMs are suppressed relative to those in (5, 2) and (5, 4) sub-nm SWNTs. These results indicate that the high curvature leads to activation of new phonon vibration in sub-nm SWNTs, which can also be inferred from the IR spectra [16]. Previously, several reports have attributed activation of weak IFMs to symmetry lowering [13]. Based on our molecular dynamics simulations and resonant Raman spectra, we believe that the high curvature-induced strain in sub-nm SWNTs along with bending of nanotubes in the bundle can severely affect some of the symmetries such as the screw axis, mirror planes and hence lead to symmetry lowering [15]. Further theoretical studies of symmetry lowering effects on vibrational modes of sub-nm SWNTs using non-periodic boundary conditions will be published elsewhere.

Next, we focus our discussion on the RBMs in Fig. 2a. We find that the resonant behavior for most of the RBMs exhibited by the samples used in this study can be adequately described using a Kataura plot with γ_0=2.45 eV [17, 18]. This value of γ_0 is in agreement with the γ_0 deduced from scanning tunneling spectroscopic (STS) studies on sub-nm SWNTs [19, 20]. For $\lambda_{excitation}$ = 785 nm, an intense RBM peak is present at 374 cm^{-1} (0.61 nm, Fig 2a). According to the original Kataura plot which uses γ_0 = 2.75 eV, E_{11} for 0.61 nm diameter tubes is ~1.28 eV and will not bring the tubes into resonance with the 785 nm (1.58 eV) excitation. Other earlier reports on ultra-small SWNTs (d ~ 0.4 nm) synthesized in zeolite crystal have also shown such failure of zone folding approximation [21, 22], where the frequency of RBM mode varies inversely with the diameter [23]. It is also worth mentioning that Heinz and co-workers [24] have reported exciton binding energies of ~400 meV for *semiconducting* 0.8 nm diameter tubes, which could bring the 0.61 nm tubes into resonance.

CONCLUSIONS

For the G-band, we observe that the S-like dispersion with respect to E_{laser} in sub-nm SWNTs tracks the electron-phonon coupling strength ($1/q$), and our molecular dymanic simulations indicate that it consists of a family of lines with $2n + m$ = constant. In this context, we note that normalizing Raman spectra with respect to the intensity of the G-band (which is does not disperse in SWNTs with $d > 1$ nm) in sub-nm SWNTs is dependent on the tube diameter. Depending on the dominant sub-SWNT diameter present in the sample, the observed intermediate frequency modes resonantly couple to specific excitation energies. Lastly, this resonance is understood in terms of curvature-induced effects in γ_0 [19].

REFERENCES

1. Jorio A, Dresselhaus MS, Dresselhaus G. *Carbon Nanotubes.* Topics in Applied Physics 111. *New York: Springer-Verlag; 2008.*
2. Samsonidze G, Saito R, Jorio A, Pimenta MA, Souza Filho AG, Gruneis A, et al. The Concept of Cutting lines in Carbon Nanotube Science. J Nanosci Nanotech 2003; **3**: 431-58.
3. Fantini C, Jorio A, Souza M, Saito R, Samsonidze, Dresselhaus MS, et al. Step-like Dispersion of the Intermediate Frequency Raman Modes in Semiconducting and Metallic Carbon Nanotubes. Phys Rev B 2005; **72**: 085446-1-5.
4. Yorikowa H, Muramastu S. Energy gaps of Semiconducting Nanotubules. Phys Rev B 1995; **52**: 2723-27.
5. Stuart SJ, Tutein AB, Harrison JA. A Reactive Potential for Hydrocarbons with Intermolecular Interactions. J Chem Phys 2000; **112**: 6472-86.

6. Li N, Wang X, Ren F, Haller GL, Pfefferle LD. Diameter Tuning of Single-Walled Carbon Nanotubes with Reaction Temperature Using a Co Mono metallic Catalyst. J Phys Chem C 2009; **113**: 10070-78.
7. Loebick CZ, Derrouiche S, Marinkovic N, Wang X, Hennrich F, Kappes M, et al. Effect of Manganese Addition to the Co-MCM-41 Catalyst in the Selective Synthesis of Single Wall Carbon Nanotubes. J Phys Chem C 2009; **113**: 21611-20.
8. Loebick CZ, Podila R, Reppert J, Chudow J, Ren F, Haller GL, et al. Selective Synthesis of Subnanometer Diameter Semiconducting Single-Walled Carbon Nanotubes. J Am. Chem. Soc 2010; **132**: 11125-31.
9. Sasaki K, Saito R, Dresselhaus G, Dresselhaus MS, Farhat H, Kong J. Curvature-induced Optical Phonon Frequency Shift in Metallic Carbon Nanotubes. Phys Rev B 2008; **77**: 245441-1-8.
10. Rao AM, Eklund PC, Bandow S, Thess A, Smalley RE. Evidence for Charge Transfer in Doped Carbon Nanotube Bundles from Raman Scattering. Nature 1997; **388**: 257-59.
11. Brown SDM, Jorio A, Corio P, Dresselhaus MS, Dresselhaus G, Saito R, et al. Origin of the Breit-Wigner-Fano Lineshape of the Tangential G-band Feature of Metallic Carbon Nanotubes. Phys Rev B 2001; **63**:155414-1-8.
12. Kwon YK, Saito S, Tomanek S. Effect of Intertube Coupling on the Electronic Structure of Carbon Nanotube Ropes. Phys Rev B 1998; **58**: R13314-17.
13. Saito R, Takeya T, Kimura T, Dresselhaus G, Dresselhaus MS. Raman Intensity of Single Walled Carbon Nanotubes. Phys Rev B 1997; **57**: 4145-53.
14. Pettifor DG, Oleinik II. Analytical bond-order potentials beyond Tersoff-Brenner. I. Theory. Phys Rev B. 1999; **59**: 8487-99.
15. Tomanek D, Enbody RJ. Science and Applications of nanotubes. New York: Kluwer Acdemic publishers; 2000.
16. Podila, R., Reppert, J., Li, N., Loebick, C.Z., Stuart, S.J., Pfefferle, L.D., Rao, A.M., Curvature-induced Symmetry Lowering and Anomalous Dispersion of the G-band in Carbon Nanotubes, Carbon (2010), doi: 10.1016/j.carbon.2010.10.033
17. Yang W, Wang RZ, Wang YF, Yan H. Are deformed modes still Raman active for single-wall carbon nanotubes? Physica B: Condensed Matter 2008; **408**: 3009-12.
18. Katuara H, Kumazawa Y, Maniwa Y, Umezu I, Suzuki S, Ohtsuka Y, et al. Optical Properties of Single Wall Carbon Nanotubes. Synthetic Metals 1999; **103**: 2555-58.
19. Odom TW, Huang JL, Kim P, Lieber CM. Atomic Structure and Electronic Properties of Single-walled Carbon Nanotubes. Nature 1998; **391**: 62-65.
20. Wildoer JWG, Venema LC, Rinzler AG, Smalley RE, Dekker C. Electronic Structure of Atomically Resolved Carbon Nanotubes. Nature 1998; **391**: 59-62.
21. Jorio A, Saito R, Hafner JH, Lieber CM, Hunter M, McClure T, et al. Structural *(n, m)* Determination of Isolated Single-wall Carbon Nanotubes by Resonant Raman Scattering. Phys Rev Lett 2001; **86**: 1118-21.
22. Kurti J, Kresse G, Kuzmany H. First-principles Calculations of the Radial Breathing Mode of Single-wall Carbon nanotubes. Phys Rev B 1998; **58**: R8869-72.
23. Tang ZK, Wang N, Zhang XX, Wang JN, Chan CT, Sheng P. Novel properties of 0.4 nm Single-walled Carbon Nanotubes Templated in the Channels of $AlPO_4$-5 single crystals. New Journal of Physics 2003; **5**:146.1-29.
24. Wang F, Dukovic G, Brus LE, Heinz TF. The Optical Resonances in carbon nanotubes Arise from Excitons. Science 2005; **308**: 838-41.

Mater. Res. Soc. Symp. Proc. Vol. 1284 © 2011 Materials Research Society
DOI: 10.1557/opl.2011.224

Properties Modeling of Low-Dimensional Carbon Nanostructures

Andrew Basteev[1], Leonid Bazyma[1], Mykhaylo Ugryumov[1], Yuriy Chernishov[1]
and Margarita Slepicheva[1]
[1]National Aerospace University, Kharkov Aviation Institute "KhAI",
17 Chkalov Str., Kharkov, 61070, Ukraine.

ABSTRACT

The modeling of single wall carbon nanotubes properties (length, diameter, chirality, defective wall structure) influence on sorption capability at different thermodynamic conditions (T= 80-273 K; P = 2-12 MPa) is presented in this work. The applied simulation procedure is the molecular dynamics as well the new event-driven simulation algorithm has been used. In the frameworks of this event-driven simulation algorithm the modeling of structure formation for carbon nanotubes have been done with different chirality and with wall defects presence. The analysis of obtained results and their comparison with published experimental and theoretical results are performed.

INTRODUCTION

Low-dimensional carbon nanostructures properties have the great potential for application in different branches of industry. We need to achieve the understanding of correlation between these objects structure and properties and the mechanism of above properties control. It is well-known that chirality of carbon nanotubes (CNT) has the significant influence on to their properties including the sorption ones [1]. In the same time enough high indexes of hydrogen mass content at relatively low pressures obtained in experiments [2] can be linked with structure defects in nanotubes walls. These defects generated for example by means of annealing procedure create the irregular surface shape and roughness which are associated with hydrogen binding energy [2]. The experimental investigations for hydrogen sorption usually give discrepant results. In connection with this the role of numerical simulation methods for sorption/desorption processes description becomes important. There are a lot of publications devoted to numerical simulation of hydrogen sorption/desorption processes [1, 3 - 6]. Usually in simulation procedures the carbon nanotubes (CNT) are considered like the perfect sheet of carbon rolled up in the cylinder. However the experimental researches showed that CNT are far from perfect structure [2]. Different kinds of defects could change electric, sorption and mechanical properties of CNT [7]. The mechanisms of topological defects formation in graphenes layers and carbon nanostructures stay still studied insufficiently. Such defects can be born at the stages of nanomaterials growth or purification and as the result of ion or electron bombardment etc. First of all one has to build the geometrical model CNT with defects in order to conduct the modeling of sorption process. The present algorithms for formation of CNT geometrical model with chaotic redistributed Stone – Wells' defects [8, 9] were used for studying of these defects influence on to CNT mechanical properties. There is a perspective to increase the CNT sorption capability due to especial defects formation in CNT structure as have been demonstrated in previous investigations [10, 11]. Although actual mathematical models are allowing conducting the analysis of CNT properties influence on to their sorption capability the modeling of defects especially defects-vacancy is needed in additional investigations.

The studying of influence of single wall carbon nanotubes properties (length, diameter, chilarity, defects in walls surface structure) on to their sorption capability at various thermodynamical conditions are performed in this work. The new event-driven simulation algorithm has been applied for modeling [12]. The modeling of carbon nanotubes structure with different chilarity including defects-vacancies presence was conducted in the frameworks of event-driven simulation method as well.

THEORY

The method which has been used in given paper generalizes the well known model of solid spheres [13]. Each modeled particle is envisioned like two-layer sphere. The events are the collisions between of inside and outside parts of model spheres. The collisions of internal parts are similar with elastic ones of solid spheres. The collisions of external parts are accompanied by following phenomena: either internal reflection without energy losses or transferring from outside into internal part with kinetic energy increasing or transferring from inside out to external part with kinetic energy losses. The travels are rectilinear and uniform between events. It gives us grounds to perform the process of model particles coordinate's centers changes as sequence of events, occurring in discrete moments of time. Moments of events coming are graduated in the order of their increasing. The velocities for both model particles are calculating for the event taking place in the nearest moment after the current one in accordance with requirement of conservation for pulse, total energy and angular momentum in interpartial collision. For each particle took part in collision the moments of time of new coming events at which it could take part are calculating. The least of these moments of time is introduced in events queue. The simplest piecewise constant potential was used: collision energy of two particles is assumed equal to zero at the distances more then d_1; collision energy of two particles is equal to infinity at the distance less then d_0 and is equal to U between above two values (see figure 1). The presence of potential barrier is postulated for the distance d_1.

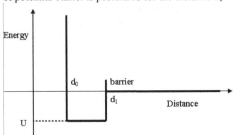

Figure 1. The rectangle potential of particles interaction.

The sequence of calculations can be described as follow:
1. The pair of spheres between which the collision has to be occurred is determined; also the moment of collision is calculated.
2. The coordinates of model spheres centers are calculated at the moment of collision.
3. The new velocities of considered pair of spheres are determined after the collision.
4. Transferring on to stage 1 till the procedure of modeling will be over.

The peculiarity of described phenomena consists in following: changes of states are occurring in discrete moments of time. The duration of time interval between two events is the value calculated at stage of current event processing. The "local object's time" is considering besides the real moment of time for following event in which the given object takes part. This time is assumed equal to moment of time t when the event happened. The acts of model objects interaction are suggested as events realizing in certain consequence obtained by means of interaction moments of time regulating in rising order. The applied calculation scheme is more effective in comparison with calculation algorithms using the constant time interval for discretization. The system evolution as whole is traced in online regime during the simulation procedure. At each step of algorithm the time is coincided with global leading parameter which is increasing discretely from event to event with intervals equivalent to interval duration between the consequent events.

The Berendsen's thermostat [14] was used in order to keep the model temperature at given level. The periodic boundary conditions were used on the computating region boundaries. The properties of interaction potentials (table I) were taken from [1]. The potential barrier for H_2-H_2 was increased for high pressures (14 MPa) which stimulate the molecular hydrogen density.

Table I. The interaction potentials barriers.

Interaction	d_0 (Å)	d_1 (Å)	U (кJ/mole)	Potential barrier U_b (кJ/mole)
$C-H_2$	3.179	3.279	2.66	0.52
H_2-H_2	2.92	3.02	3.06	1.80

DISCUSSION

The method verification

The comparative calculations were done in the frameworks of Event-Driven Simulation and HyperChem 8.0 with the aim to verify of new algorithm. Calculations in HyperChem 8.0 were conducted by means of molecular dynamics method for force field Amber94. The target setting was taken from [11] for carbon nanotubes "armchair" type with chilarity (8,8), (10,10), (12,12) at temperature 80 K and pressure 14 MPa. The computation region in [1] was constant 40×80×40 Å but in our calculations it was varied in some limits (table II) for different CNT (D_{CNT} - diameter; L_{CNT} - length; N(C) - number of carbon atoms).

Table II. CNT geometrical properties.

Chilarity	Type	D_{CNT} (Å)	L_{CNT} (Å)	N(C)	Calculation Area (Å)
(12, 12)	"armchair"	16.28	38.13	768	56×56×56
(10, 10)	"armchair"	13.56	38.13	640	51×56×51
(8, 8)	"armchair"	10.84	38.13	512	46×56×46

The maximum time for dynamical equilibrium achievement (1015 ps) was observed for CNT (12,12). The mistake analysis showed that this is inversely to square root from number of associated with CNT molecules. For the mass hydrogen content (table III) the mistake is within the interval ± (0.15-0.18) %.

Table III. Hydrogen sorption capability for CNT "armchair" type (T=80 K, P =14 MPa).

Chilarity	H₂, (outside CNT)			H₂, (inside CNT)			H₂ mass content (%)		
	Event Chem	Hyper Chem	[1]	Event Chem	Hyper Chem	[1]	Event Chem	Hyper Chem	[1]
(12, 12)	331	263	329	115	163	112	9.68	9.24	9.57
(10, 10)	255	215	301	94	108	66	9.08	8.41	9.56
(8, 8)	236	222	271	73	72	43	10.05	9.56	10.22

It is clear that there is the satisfactory fit of total mass hydrogen content obtained with use of HyperChem, EventChem and taken from [1]. In the same time there is any discrepancy in hydrogen molecules redistribution on the CNT internal and external surfaces.

CNT diameter influence

In order to study diameter influence on searched characteristics the CNT with chirality (6,6) with following geometrical properties D_{CNT}= 8.14 Å; L_{CNT}= 38.13 Å; N(C)=384 and calculation region sizes 40×56×40 Å for P = 14 MPa and 85×85×85 Å for P = 2 MPa were considered additionally to above studied objects. It should be noted that at pressure changes the sizes of calculation regions changed as well for all types of CNT. It is necessary to note that the mistake in hydrogen mass content determination in CNT with minimal diameter will be more than one for other CNT (table III). The mistake is equal to ±0.22 percent and at P = 2 MPa mistake will be maximal for all considered CNT i.e. ±0.13 percent. It is well known that CNT specific surface is increasing with diameter increasing and high value of hydrogen mass capability should be expected for big diameters. However the obtained data allow saying about this tendency at relatively low pressures P = 2 MPa only (table IV).

Table IV. Hydrogen sorption capability for CNT "armchair" type for different diameters.

Chilarity	H₂, (outside CNT)		H₂, (inside CNT)		H₂ mass content (%)	
	14 MPa	2 MPa	14 MPa	2 MPa	14 MPa	2 MPa
(12, 12)	331	129	115	74	9.68	4.21
(10, 10)	255	127	94	48	9.08	4.55
(8, 8)	236	99	73	27	10.05	4.10
(6, 6)	214	75	18	8	10.07	3.61

CNT type influence

CNT with different type structure and clear diameters were studied in this part of the work: "armchair" type with chirality (7,7) D_{CNT}= 9.5 Å; L_{CNT}= 39.36 Å; N(C)=462; "zigzag" type with chrality (12,0), D_{CNT}= 9.38 Å, L_{CNT}= 38.69 Å, N(C)=456; "armchair" type with chirality (8, 8), D_{CNT}= 10.84 Å, L_{CNT}= 38.13 Å, N(C)=512 and "zigzag" type with chirality (14, 0), D_{CNT}= 10.97 Å, L_{CNT}= 38.69 Å; N(C)=532. The mistake in estimation of mass hydrogen content is approximately the same for all considered CNT. From table V we may conclude that for CNT with clear diameters and equal length the hydrogen mass content depends upon CNT type. For CNT with "zigzag" type this content is more at the same other conditions. This fact is in accordance with conclusions made in [9].

Table V. Hydrogen sorption capability for different types CNT (T= 80 K; P=2 MPa).

Chilarity	Type	H$_2$, (outside CNT)	H$_2$, (inside CNT)	H$_2$ mass content (%)
(7,7)	"armchair"	71	13	3.03±0.11
(12,0)	zigzag"	92	12	3.80±0.12
(8,8)	"armchair"	99	27	4.10±0.12
(14,0)	"zigzag"	122	24	4.57±0.13

The CNT wall defects influence

The influence of "armchair" CNT wall defects (vacancies) and chilarity (10,10) in calculation region 101×101×101 Å at T=80 K and P=2 MPa was analyzed in this part. The level of defects saturation was chosen in the interval (0.8…13.3) percent. Taking into account approximately same mistake in obtained results we have to note the increasing of hydrogen mass content with increasing of defects-vacancies number (table VI). However at relatively high level of CNT defects penetration the hydrogen mass content increasing is linked due to CNT mass decreasing.

Table VI. Hydrogen sorption for CNT with defects.

Vacancies (%)	H$_2$, (outside CNT)	H$_2$, (inside CNT)	H$_2$ mass content (%)
0 (0%)	127	48	4.32±0.11
5 (0.8%)	143	48	5.01±0.12
18 (2.8%)	146	49	5.22±0.12
67 (10.5%)	140	61	5.84±0.14
85 (13.3%)	138	44	5.56±0.13

Pressure and temperature influence

This part is devoted to modeling of pressure and temperature influence on to CNT (12, 12), hydrogen mass sorption capability for pressure (P = 2-14 MPa) at T=80 K and at T = 273 K obtained by means of EventChem (see figure 2).

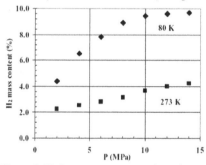

Figure 2. Hydrogen mass content dependence upon pressure.

It has been discovered that at low temperatures two layers of adsorbed molecules can be formed. The density of the second layer of adsorbed molecules is sufficiently lower than the density of the first layer but it significantly increases the total hydrogen mass content due to big diameter. The formation of the second adsorption layer was not observed at temperature T = 273 K. The mass hydrogen content published in [15] for individual CNT at room temperature is in good accordance with results obtained by means of EventChem.

CONCLUSIONS

The results achieved with use of Event-Driven Simulation are numerically and qualitatively in accordance with other methods of molecular dynamics. The mistake of results is estimated by the value which is proportional to square root from number of molecules associated with CNT. The CNT diameter influence on hydrogen mass content in CNT is effective at relatively low pressures (~2 MPa). The hydrogen mass content in CNT with clear diameters and equal length is defined by CNT type. For CNT "zigzag" type the content is more on (0.5-0.8) % than for CNT "armchair" type. The generation of defects-vacancies in CNT structure is the reason of relative hydrogen mass content increasing absorbed in CNT. The saturation of CNT by defects at level (0.8-13.3) % this increasing is about (0.7-1.2) %. Two layers of adsorbed molecules could be formed at low temperatures. The density of the second layer of adsorbed molecules is sufficiently lower than the density of the first layer but the second layer significantly increases the total hydrogen mass content due to big diameter. The formation of the second adsorption layer was not observed at temperature T = 273 K.

ACKNOWLEDGMENTS

Performed investigation was encouraged by Science and Technology Center in Ukraine, Project # 4957.

REFERENCES

1. GUO Lianquan, MA. Changxiang, Wang Shuai, MA He and LI Xin, *J. Mater. Sci. Technol.* **21**, 123 (2005).
2. B.K. Pradhan, A.R. Harutyunyan, D. Stojkovic, J.C. Grossman, P. Zhang, M.W. Cole, V. Crespi, H. Goto, J. Fujiwara and P.C. Eklund, *J. Mater. Res.* **17**, 2209 (2002).
3. H. Cheng, A.C. Cooper, G.P. Pez, M.K. Kostov, P. Piotrowski and S.J. Stuart, *J. Phys. Chem. B.* **109**, 3780 (2005).
4. L.Verlet, *Phys. Rev.* **159**, 103 (1967).
5. F. Huarte-Larranaga and M. Alberti, *Chem. Phys. Lett.* **445**, 227 (2007).
6. Yuchen Ma, Yueyuan Xia, Mingwen Zhao and Minju Ying, *Phys. Rev. B.* **65**, 155430 (2002).
7. M.A. Obolensky, A.V. Basteev and L.A. Bazyma, *Fullerenes, Nanotubes, and Carbon Nanostructures* **19**, (2010) (in press)
8. L.G. Zhou and San-Qiang Shi, *App. Phys. Lett.* **83**, 1222 (2003).
9. Qiang Lu and Baidurya Bhattacharya, *Nanotechnology* **16**, 555 (2005).
10. Wen Qian, Jiafu Chen, Lingzhi Wei, Liusuo Wu and Qianwang Chen, *NANO* **4**, 7 (2009).
11. L.G. Zhou, F.Y. Meng and S.Q. Shi, *Nanotechnology* **3**, 154 (2003).
12. Yu.C. Chernyshev and M.A. Slepicheva, Ukraine inventor's certificate No. 34404 (5 August 2010)
13. B.J. Alder and T.E. Wainwright, *J. Chem. Phys.* **27**, (1957).
14. H. J. Berendsen, *J. Chem. Phys.* **81**, 3684 (1984).
15. N. Hu, X. Sun and A. Hsu, *J. Chem. Phys.* **123**, 044408 (2005).

Doping, Defects and Surface Chemistry

Mater. Res. Soc. Symp. Proc. Vol. 1284 © 2011 Materials Research Society
DOI: 10.1557/opl.2011.225

Effect of Sidewall Fluorination on the Mechanical Properties of Catalytically Grown Multi-Wall Carbon Nanotubes

Yogeeswaran Ganesan[1], Cheng Peng[1], Lijie Ci[1], Valery Khabashesku[2], Pulickel M. Ajayan[1] and Jun Lou[1]

[1] Department of Mechanical Engineering and Materials Science, Rice University, 6100 Main Street, Houston, TX, 77006
[2] Department of Chemical and Biomolecular Engineering, University of Houston, 4800 Calhoun Road, Houston, Texas, 77004

ABSTRACT

We report on the usage of a simple microfabricateddevice, that works in conjunction with a quantitative nanoindenter inside a scanning electron microscope (SEM), for the *in situ* quantitative tensile testing of individual sidewall fluorinated multi-wall carbon nanotubes (MWNTs). The stress vs. strain curves and the tensile strength values for five fluorinated specimens have been presented and compared to those of pristine MWNT specimens (data reported earlier). The fluorinated specimens were found to deform and fail in a brittle fashion similar to pristine MWNTs. However, sidewall fluorination was found to have considerably degraded the mechanical properties (tensile strength and load bearing capacity) of the MWNTs.

INTRODUCTION

The preparation, processing, and property tuning of carbon nanotubes (CNTs) reinforced nanocomposites require the dispersion and solubilization of CNTs, which in their pristine form are not soluble in most common organic solvents and water. Chemical modification of carbon nanotubes with functional groups has been found to be an excellent method to promote dispersion (by de-bundling) and also to improve their interaction with a matrix material via hydrogen or covalent bonding. In recent years, several approaches to achieve the functionalization of carbon nanotubes have been developed, in both molecular and supramolecular chemistry. These approaches include defect functionalization[1], covalent functionalization of the side-walls[2], noncovalentexohedral functionalization[3] andendohedral functionalization [4]. Besides a general improvement in the solubility and processibility that can be achieved by all these approaches, sidewall functionalizations are particularly interesting since they significantly alter the structural and electronic properties of carbon nanotubes, yielding new nanotube derivatives with useful properties of their own [5]. However, modifying the nanotubes by sidewall functionalization changes the surface structure since it results in the cleavage of carbon-carbon bonds along the graphite sidewall, therefore degrading their intrinsic mechanical properties.

The direct addition of fluorine [5], hydrogen [6], aryl groups [7,8], nitrenes, carbenes, and radicals [9,10]among others, to the side walls of pristine SWNTs been reported in the past. Fluorination as a covalent functionalization strategy is considered particularly important since it can improve dispersion considerably and because fluorine can be substituted with more complex addends, opening the way to more complex chemical functionalization of nanotubes for improved covalent interactions with matrix materials [11]. In the earliest reports on sidewall functionalization chemistry, it was shown that fluorine substituents on SWNTs can be substituted by alkyl groups from corresponding Grignard and alkyllithium reagents, resulting in the covalent attachment of alkyls to the nanotube sidewalls through the C-C bonds [5]. These reactions are facilitated by weakened C-F bonds relative to those in alkylfluorides and a stronger electron-accepting ability of fluoronanotubes in comparison with that of pristine carbon nanotubes. In addition to this, partial removal of functional groups from the surface of fluoronanotubes during processing with an epoxy matrix has been observed in the past, suggesting that fluorination could itself facilitate *in situ* direct covalent bonding between nanotubes and a matrix material, ultimately resulting in mechanical reinforcement of the composite [11]. Thus far, no systematic experimental data can be found in literature that discusses the effects of fluorination on mechanical properties of CNTs. Hence, the novel nanoindenter assisted technique (see Fig. 1), described in detail earlier [12], was used to probe the mechanical properties of individual fluorinated MWNTs (F-MWNTs) by uniaxial tensile testing, *in situ*, within an SEM chamber.

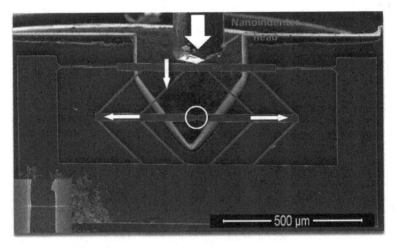

Figure1. SEM image of the novel microfabricated device; block arrows show the direction of movement of the indenter tip and the shuttles during the experiment; (Inset) high magnification image of circled region shows a nanotube specimen across the sample stage shuttle gap

EXPERIMENT

Pristine MWNTs specimens were grown on bare quartz substrates by injecting a mixture of 20mg/ml of ferrocene ($(C_5H_5)_2Fe$) in xylene (C_8H_{10}) solution into a two-stage thermal CVD reactor consisting of a low temperature (200 °C) pre-heater followed by a higher temperature main reactor (775 °C). A Mixture of 10 % elemental fluorine and 90% helium was used as the fluorinating agent for the MWNTs. This mixture along with additional helium gas feed was passed through a temperature controlled Monel flow reactor, held at 160 °C, containing the nanotube sample. A 4% increase in the weight of the samples was observed upon fluorination. X ray photoelectron spectroscopy was conducted on the MWNTs (data not shown) in order to determine the C:F ratio on the surface of the MWNTs viz. 77.9:22.1. Note that transmission electron microscopy (TEM) images showed no significant changes in the morphology of the MWNTs upon fluorination (images not shown).

The micro-fabricated device used to perform the tensile experimentsconsists of a pair of movable (sample stage) shuttles that are attached to a top shuttle via inclined freestanding beams (see Fig. 1). Its actuation involves the usage of a nanoindenter that applies a load on the top shuttle of the device along the y axis; four sets of inclined symmetrical beams transform the motion of the top shuttle into a two dimensional translation of the sample stage shuttles and ensures that the stress applied on a specimen mounted across the sample stage shuttles is purely tensile. The use of the device is advantageous since it allows the application and measurement of forces with nano-Newton resolution and measurement of local mechanical deformation with nanometer resolution. This is because the force and displacement resolution of the devices are dictated by that of the nanoindenter viz. 69.4 nano-newtons and 8.675 angstroms respectively. The devices were fabricated on SOI wafers using standard photolithographic techniques. Details regarding sample mounting and indentation experiments can be found in an earlier manuscript [13].

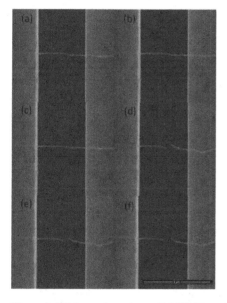

Figure 2. SEM snapshots show a F-MWNT specimen undergoing deformation under a tensile load at (a) t=0, (b) t= 4, (c) t= 8, (d) t= 9, (e) t= 12 and (f) t= 18 seconds

DISCUSSION

The stress vs. strain curves for the MWNT samples, shown in Fig. 3, were extracted from their corresponding indenter load vs. displacement curves via a response subtraction technique and were found to be almost always linear prior to fracture. When defect free MWNTs are subject to tensile loading, only the outermost wall of each tube are considered to be load bearing. Such MWNTs fail via a "sword in sheath" mechanism with the inner walls experiencing a pullout after failure of the outermost wall [14]. In the case of the catalytically grown MWNTs though, the existence of a large density of defects leads to inter-shell cross-linking as a result of which multiple graphitic walls bear tensile loads. Hence, when the MWNTs fail, multiple wall fracture can be observed at the point of nanotube failure. Significant intershell cross-linking between the graphitic shells in the catalytically grown pristine MWNTs was found to result in load sharing in a fashion that caused all graphitic shells to fracture in close proximity to one another, at the point of failure [13]. SEM images of the fluorinated MWNTs tested, post failure, revealed similar flat (if somewhat corrugated) post failure surfaces (see Fig. 4); stress vs. strain curves were thus plotted assuming that the entire cross-section area of each fluorinated MWNT was load bearing (Fig. 3). The average strength and maximum load borne values (1.026 GPa and 6.35 μN) were found to be much lower that of pristine MWNTs (2.134 GPa and 16.495 μN) (see Table 1).

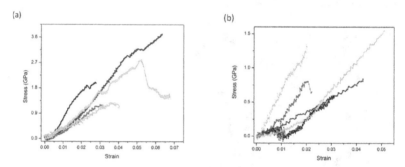

Figure 3. Plots show the stress vs. strain curves for (a) 5 pristine MWNTs and (b) 5 sidewall fluorinated MWNTs

In defect free MWNTs, sidewall fluorination would result in the formation of C-F bonds (or defects) only on the outermost shell of the nanotubes. Hence, considerable degradation of mechanical properties would be expected to occur upon fluorination, in the case of initially defect free MWNTs without significant intershell crosslinking [15]. With regard to the catalytically grown MWNTs, known to possess high defect densities, fluorine incorporation could occur on more than one outmost graphitic shell of each nanotube. Thus, considerable changes in the mechanical properties would also be expected upon fluorination of such tubes. Comparison of the strength and maximum load borne values with those obtained by testing pristine MWNTs suggest that (a) significant degradation in the mechanical properties of catalytically grown MWNTs does occur upon sidewall fluorination and that (b) while the entire cross-section area of the catalytically grown MWNTs can be considered load-bearing owing to defect based cross-linking [13], load distribution among the shells is likely to be non-uniform, with the bulk of the tensile loads being borne by the outermost shells.

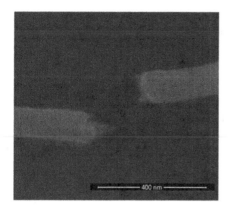

Figure 4. SEM image shows fracture surface of a sidewall fluorinated MWNT

Table 1. Mechanical properties of Pristine and Fluorinated MWNTs as ascertained from the tensile tests

MWNT type	Maximum Load (nN)	Tensile strength (GPa)
Pristine	6873	1.20
Pristine	36838	3.72
Pristine	15929	1.96
Pristine	11179	0.99
Pristine	11654	2.80
Fluorinated	4631	1.34
Fluorinated	4906	0.80
Fluorinated	11045	1.55
Fluorinated	7677	0.84
Fluorinated	3494	0.60

CONCLUSIONS

The mechanical properties of sidewall fluorinated catalytically grown MWNTs were characterized by *in situ* tensile testing within an SEM chamber using a novel nanoindenterassisted technique. The average tensile strength value of the MWNTs, computed assuming that all the walls of the MWNTs were load bearing, was found to be considerably lower (by about 50%) that that for pristine MWNTs. Thus, while on one hand, sidewall functionalization techniques can potentially improve interfacial adhesion in MWNT based composites, they can, at the same time, significantly reduce the mechanical strength of the MWNTs themselves.

ACKNOWLEDGMENTS

This work was supported by the National Science foundation CMMI 0800896, the Welch Foundation grant C-1716, the Air Force Research laboratory grant FA8650-07-2-5061 and the Air Force Office of Sponsored Research Young Investigator Program award FA9550-09-1-0084. The authors would like to thank Dr. Yang Lu and Dr. Yongjie Zhan for assistance with transmission electron microscopy.

REFERENCES

1. M. A. Hamon, H. Hu, P. Bhowmik, S. Niyogi, B. Zhao, M. E. Itkis, R. C. Haddon, Chem. Phys. Lett., 347, 8 (2001).
2. J. L. Stevens, A. Y. Huang, H. Peng, I. W. Chiang, V. N. Khabashesku, J. L. Margrave, Nano Lett., 3, 331 (2003).
3. F. Balavoine, P. Schultz, C. Richard, V. Mallouh, T. W. Ebbesen, C. Mioskowski, Angew. Chem., 111, 2036 (1999).
4. B. W. Smith, M. Monthioux, D. E. Luzzi, Chem. Phys. Lett., 315, 31 (1999).
5. E. T. Mickelson, C. B. Huffman, A. G. Rinzler, R. E. Smalley, R. H. Hauge, J. L. Margrave, Chem. Phys. Lett., 296, 188 (1998).
6. S. Pekker, J.-P.Salvelat, E. Jakab, J.-M. Bonard, L. J. Forro, Phys. Chem. B, 105, 7938 (2001).
7. J. L. Bahr, J. Yang, D. V. Kosynkin, M. J. Bronikowski, R. E. Smalley, J. M. Tour, J. Am. Chem. Soc., 123, 6536 (2001).
8. J. L. Bahr, J. M. Tour, Chem. Mater., 13, 3823 (2001).
9. M. Holzinger, O. Vostrowsky, A. Hirsch, F. Hennrich, M. Kappes, R. Weiss, F. Jellen, Angew. Chem., Int. Ed., 40, 4002 (2001).
10. H. Peng, P. Reverdy, V. N. Khabashesku, J. L. Margrave, Chem. Commun., 362 (2003).
11. J. Zhu, J. Kim, H. Peng, J. L. Margrave, V. N. Khabashesku, E. V. Barrera, Nano Lett., 3, 1107 (2003).
12. Y. Ganesan, Y. Lu, C. Peng, H. Lu, R. Ballarini and J. Lou, JMEMS, 19, 675 (2010).
13. Y. Ganesan, C. Peng, Y. Lu, L. Ci, A. Srivastava, P. M. Ajayan and J. Lou, ACS Nano, Published online (10.1021/nn102372w) (2010).
14. M.F. Yu, O. Lourie, M.J. Dyer, K. Moloni, T.F. Kelly and R.S. Ruoff, Science 287,637(2000).
15. R. Khare, S. L. Mielke, J. T. Paci, S. Zhang, R. Ballarini, G. C. Schatz, T. Belytschko, Phys. Rev. Lett., 75, 075142 (2007).

Novel Structures and Properties of Low Dimensional Carbon Nanostructures

Mater. Res. Soc. Symp. Proc. Vol. 1284 © 2011 Materials Research Society
DOI: 10.1557/opl.2011.226

Nonlinear Mechanical Properties of Graphene Nanoribbons

Qiang Lu and Rui Huang
Department of Aerospace Engineering and Engineering Mechanics, University of Texas, Austin, Texas 78712, USA

ABSTRACT

Based on atomistic simulations, the nonlinear elastic properties of monolayer graphene nanoribbons under quasistatic uniaxial tension are predicted, emphasizing the effect of edge structures (armchair and zigzag, without and with hydrogen passivation). The results of atomistic simulations are interpreted using a theoretical model of thermodynamics, which enables determination of the nonlinear functions for the strain-dependent edge energy and the hydrogen adsorption energy, for both zigzag and armchair edges. Due to the edge effects, the initial Young's modulus of graphene nanoribbons under infinitesimal strain varies with the edge chirality and the ribbon width. Furthermore, it is found that the nominal strain to fracture is considerably lower for armchair graphene nanoribbons than for zigzag ribbons. Two distinct fracture mechanisms are identified, with homogeneous nucleation for zigzag ribbons and edge-controlled heterogeneous nucleation for armchair ribbons.

INTRODUCTION

Graphene ribbons with nanoscale widths ($W < 20$ nm) have been produced recently, either by lithographic patterning [1-3] or by chemically derived self assembly processes [4], with potential applications in nanoelectronics and electromechanical systems. The edges of the graphene nanoribbons (GNRs) could be zigzag, armchair, or a mixture of both [5]. It has been theoretically predicted that the special characteristics of the edge states leads to a size effect in the electronic state of graphene and controls whether the GNR is metallic, insulating, or semiconducting [5-8]. The effects of the edge structures on deformation and mechanical properties of GNRs have also been studied to some extent [9-18]. On one hand, elastic deformation of GNRs has been suggested as a viable method to tune the electronic structure and transport characteristics in graphene-based devices [15, 16]. On the other hand, plastic deformation and fracture of graphene may pose a fundamental limit for reliability of integrated graphene structures.

The mechanical properties of bulk graphene (i.e., infinite lattice without edges) have been studied both theoretically [19-21] and experimentally [22]. For GNRs, however, various edge structures are possible [23, 24], with intricate effects on the mechanical properties. Ideally, the mechanical properties of GNRs may be characterized experimentally by uniaxial tension tests. To date however no such experiment has been reported, although similar tests were performed for carbon nanotubes (CNTs) [25]. Theoretically, previous studies on the mechanical properties of GNRs have largely focused on the linear elastic properties (e.g., Young's modulus and Poisson's ratio) [11-15]. While a few studies have touched upon the nonlinear mechanical behavior including fracture of GNRs [12, 13, 16], the effect of the edge structures in the nonlinear regime has not been well understood. In the present study, based on atomistic simulations and a thermodynamics model, the nonlinear elastic deformation of graphene nanoribbons under quasistatic uniaxial tension are analyzed, emphasizing the effects of edge structures on elastic modulus and fracture strength.

ATOMISTIC SIMULATIONS

Atomistic simulations of GNRs under uniaxial tension are performed using the second-generation reactive empirical bond order (REBO) potential [26]. In each simulation, the tensile strain is applied incrementally in the longitudinal direction of the GNR, until fracture occurs. At each strain level, the statically equilibrium lattice structure of the GNR is calculated to minimize the total potential energy by a quasi-Newton algorithm. Periodic boundary conditions are applied at both ends of the GNR, whereas the two parallel edges (zigzag or armchair) of the GNR are free of external constraint. To study the effect of hydrogen passivation along the free edges, the results for GNRs with bare and passivated edges are compared. The mechanical behavior of infinite graphene lattice is also simulated by applying the periodic boundary conditions at all four edges, for which the uniaxial stress state is achieved by lateral relaxation perpendicular to the loading direction.

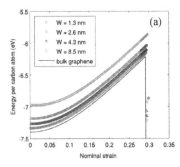 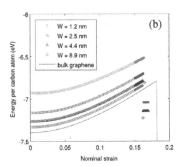

Figure 1. Potential energy per carbon atom as a function of the nominal strain for graphene nanoribbons under uniaxial tension, with (a) zigzag and (b) armchair edges, both unpassivated.

Figure 1 shows the results from atomistic simulations for GNRs with unpassivated edges, where the ribbon width (W) is varied between 1 and 10 nm. For each GNR, the average potential energy per carbon atom increases as the nominal strain increases until it fractures at a critical strain. To understand the numerical results, we adopt a simple thermodynamics model for the uniaxially stressed GNRs. For a GNR of width W and length L, the total potential energy as a function of the nominal strain consists of contributions from deformation of the interior lattice (bulk strain energy) and from the edges (edge energy), namely

$$\phi(\varepsilon) = \phi_0 WL + U(\varepsilon)WL + 2\gamma(\varepsilon)L, \qquad (1)$$

where ε is the nominal strain (relative to the bulk graphene lattice at the ground state), ϕ_0 is the potential energy density (per unit area) of graphene at the ground state, $U(\varepsilon)$ is the bulk strain energy density (per unit area), and $\gamma(\varepsilon)$ is the edge energy density (per unit length of the edges).

While the bulk strain energy density as a function of the nominal strain can be obtained directly from the atomistic calculations for the infinite graphene lattice (dashed lines in Fig. 1), the edge energy density function is determined by subtracting the bulk energy from the total potential energy of the GNRs based on Eq. (1). Thus, both the energy functions are atomistically determined, which can then be fitted with nonlinear polynomial functions for theoretical purposes [20].

The GNR under uniaxial tension is subjected to a net force (F) in the longitudinal direction. At each strain increment, the mechanical work done by the longitudinal force equals the increase of the total potential energy, which can be written in a variational form, i.e.,

$$\delta\phi = FL\delta\varepsilon . \tag{2}$$

Consequently, the force (F) can be obtained from the derivative of the potential energy function in Eq. (1). A two-dimensional (2-D) nominal stress can then be defined without ambiguity as the force per unit width of the GNR, namely

$$\sigma(\varepsilon) = \frac{F}{W} = \frac{dU}{d\varepsilon} + \frac{2}{W}\frac{d\gamma}{d\varepsilon}. \tag{3}$$

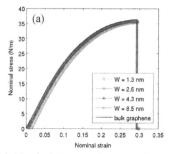
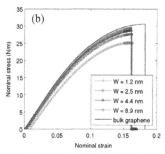

Figure 2. Nominal stress-strain curves for graphene nanoribbons under uniaxial tension, with (a) zigzag and (b) armchair edges, both unpassivated.

Figure 2 shows the 2-D nominal stress-strain curves of the GNRs obtained by taking the derivative of the potential energy curves in Fig. 1. Apparently, the stress-strain relation of a GNR is generally nonlinear, for which the tangent modulus as a function of the nominal strain is defined as

$$E(\varepsilon) = \frac{d\sigma}{d\varepsilon} = \frac{d^2U}{d\varepsilon^2} + \frac{2}{W}\frac{d^2\gamma}{d\varepsilon^2}. \tag{4}$$

The first term on the right-hand side of Eq. (4) represents the tangent modulus of the bulk graphene lattice (under the condition of uniaxial stress), and the second term is the contribution from the edges (i.e., edge modulus). Therefore, the elastic modulus of the GNR in general depends on the ribbon width (W) as well as the edge chirality. Similar stress-strain curves were obtained by molecular dynamics (MD) simulations [12], where the critical strain to fracture is typically lower than the static MM simulations due to the effects of temperature and loading rate.

DISCUSSIONS

The nominal stress-strain curves in Fig. 2 show approximately linear elastic behavior of all GNRs at relatively small strains (e.g., $\varepsilon < 5\%$). Following Eq. (4), the initial Young's modulus of the GNRs in the linear regime can be written as

$$E_0 = E_0^b + \frac{2}{W} E_0^e,$$ (5)

where E_0^b is the initial Young's modulus of the bulk graphene and E_0^e is the initial edge modulus. While the bulk graphene is isotropic in the regime of linear elasticity, the edge modulus depends on the edge chirality with different values for the zigzag and armchair edges. As a result, the initial Young's modulus of the GNR depends on the edge chirality and the ribbon width, as shown in Fig. 3. The REBO potential used in the present study predicts a bulk Young's modulus, $E_0^b = 243$ N/m, and the predicted edge modulus is $E_0^e = 8.33$ nN (~52 eV/nm) for the unpassivated zigzag edge and $E_0^e = 3.65$ nN (~22.8 eV/nm) for the unpassivated armchair edge. With positive moduli for both edges, the Young's modulus of unpassivated GNRs increases as the ribbon width decreases. We note that the predicted edge modulus is considerably lower than a previous calculation using a different potential [11], and the REBO potential is known to underestimate the bulk modulus [27, 28].

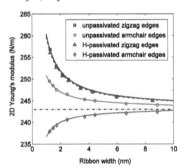

Figure 3. Initial Young's modulus versus ribbon width for GNRs with unpassivated and hydrogen-passivated edges.

For a GNR with hydrogen (H) passivated edges, the potential energy in Eq. (1) is modified to account for the hydrogen adsorption, namely

$$\phi(\varepsilon) = \phi_0 WL + U(\varepsilon)WL + 2\gamma(\varepsilon)L - 2\gamma_H(\varepsilon)L,$$ (6)

where $\gamma_H(\varepsilon)$ is the adsorption energy of hydrogen per unit length along the edges of the GNR and the negative sign indicates typically reduced edge energy due to hydrogen passivation [17, 23]. By comparing the potential energies for the GNRs with and without H-passivation, the adsorption energy can be determined as a function of the nominal strain for both armchair and zigzag edges. At zero strain ($\varepsilon = 0$), our MM calculations predict the hydrogen adsorption energy to be 20.5 and 22.6 eV/nm for the zigzag and armchair edges, respectively, which compare closely with the first-principle calculations [23]. Under uniaxial tension, the adsorption energy varies with the nominal strain. Similar to Eq. (3), the nominal stress for the H-passivated GNR is obtained as

$$\sigma(\varepsilon) = \frac{dU}{d\varepsilon} + \frac{2}{W}\left(\frac{d\gamma}{d\varepsilon} - \frac{d\gamma_H}{d\varepsilon}\right). \tag{7}$$

The effect of hydrogen passivation on the initial Young's modulus of GNRs is shown in Fig. 3. Interestingly, while hydrogen passivation has negligible effect on the initial Young's modulus of GNRs with zigzag edges, the effect is dramatic for GNRs with armchair edges. In the latter case, a negative edge modulus ($E_0^e = -20.5 \, \text{eV/nm}$) is obtained, and thus the initial Young's modulus decreases with decreasing ribbon width, opposite to the unpassivated GNRs.

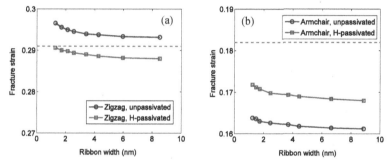

Figure 4. Fracture strain versus ribbon width for GNRs under uniaxial tension, with (a) zigzag and (b) armchair edges. The horizontal dashed line in each figure indicates the fracture strain of bulk graphene under uniaxial tension in the same direction.

Without any defect, the bulk graphene fractures when the tangent modulus becomes zero (i.e., $d^2U/d\varepsilon^2 = 0$), dictated by the intrinsic lattice instability under tension [19-21]. At a finite temperature, however, fracture may occur much earlier due to thermally activated processes [12]. As shown in a previous study [21], the critical strain to fracture for bulk graphene varies with the loading direction. Both first-principle calculations [19, 20] and empirical potential models [21] have predicted that the intrinsic critical strain is higher for graphene under uniaxial tension in the zigzag direction than in the armchair direction, suggesting that the hexagonal lattice of graphene preferably fractures along the zigzag directions by cleavage. As shown in Fig. 1a, the GNRs with zigzag edges fracture at a critical strain close to that of bulk graphene loaded in the same direction. In contrast, Fig. 1b shows that the GNRs with armchair edges fracture at a critical strain considerably lower than bulk graphene. In both cases, the fracture strain slightly depends on the ribbon width, as shown in Fig. 4. Hydrogen passivation of the edges leads to slightly lower fracture strains for zigzag GNRs, but slightly higher fracture strains for armchair GNRs.

The apparently different edge effects on the fracture strain imply different fracture mechanisms for the zigzag and armchair GNRs. The processes of fracture nucleation in GNRs are studied by molecular dynamics (MD) simulations at different temperatures. It is found that the edge effect leads to two distinct mechanisms for fracture nucleation in GNRs at relatively low temperatures ($T < 300$ K). Figure 5 shows two fractured GNRs at 50 K. For the GNR with zigzag edges (Fig. 5a), fracture nucleation occurs stochastically at the interior lattice of the zigzag GNRs. As a result, the fracture strain is very close to that of bulk graphene strained in the same direction, consistent with the MM calculations. However, for the GNR with armchair edges (Fig. 5b), fracture nucleation occurs exclusively near the edges. Thus, the armchair edge serves as the preferred location for fracture nucleation, leading to a lower fracture strain compared to bulk graphene, as seen also from the MM calculations. Therefore, two distinct fracture nucleation mechanisms are identified as interior homogeneous nucleation for the zigzag GNRs and edge-controlled heterogeneous nucleation for the armchair GNRs. The same mechanisms hold for GNRs with H-passivated edges. It is evident from Fig. 5 that cracks preferably grow along the zigzag directions of the graphene lattice in both cases.

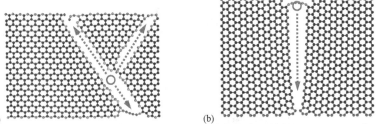

(a)   (b)

Figure 5. Fracture of graphene nanoribbons under uniaxial tension. (a) Homogeneous nucleation for a zigzag GNR; (b) edge-controlled heterogeneous nucleation for a armchair GNR. The circles indicate the nucleation sites, and the arrows indicate the directions of crack growth.

CONCLUSIONS

In summary, this paper presents a theoretical study on the effects of edge structures on the mechanical properties of graphene nanoribbons under uniaxial tension. Due to the edge effect, the initial Young's modulus of GNRs under infinitesimal strain depends on both the chirality and the ribbon width. Furthermore, it is found that the strain to fracture is considerably lower for armchair GNRs than that for zigzag GNRs. Two distinct fracture mechanisms are identified, with homogeneous nucleation for the zigzag GNRs and edge-controlled heterogeneous nucleation for the armchair GNRs. Hydrogen passivation is found to have relatively small effects on the mechanical behavior of zigzag graphene ribbons, but its effect is more significant for armchair ribbons.

ACKNOWLEDGMENTS

The authors gratefully acknowledge funding of this work by National Science Foundation through Grant No. 0926851.

REFERENCES

1. C. Berger, et al., *Science* **312**, 1191-1196 (2006).
2. B. Ozyilmaz, et al., *Phys. Rev. Lett.* **99**, 166804 (2007).
3. Y. M. Lin, et al., *Phys. Rev. B* **78**, 161409R (2008).
4. X. L. Li, et al., *Science* **319**, 1229-1232 (2008).
5. K. Nakada, M. Fujita, G. Dresselhaus, M. S. Dresselhaus, *Phys. Rev. B* **54**, 17954-17961 (1996).
6. Y.-W. Son, M. L. Cohen, S. G. Louie, *Phys. Rev. Lett.* **97**, 216803 (2006).
7. V. Barone, O. Hod, G. E. Scuseria, *Nano Lett.* **6**, 2748-2754 (2006).
8. S. Dutta, S. Lakshmi, S. K. Pati, *Phys. Rev. B* **77**, 073412 (2008).
9. V. B. Shenoy, C. D. Reddy, A. Ramasubramaniam, Y. W. Zhang, *Phys. Rev. Lett.* **101**, 245501 (2008).
10. K. V. Bets, B. I. Yakobson, *Nano Research* **2**, 161-166 (2009).
11. C. D. Reddy, A. Ramasubramaniam, V. B. Shenoy, Y. W. Zhang, *Appl. Phys. Lett.* **94**, 101904 (2009).
12. H. Zhao, K. Min, N. R. Aluru, *Nano Lett.* **9**, 3012-3015 (2009).
13. H. Bu, et al., *Phys. Lett. A* **373**, 3359-3362 (2009).
14. Z. P. Xu, *J. Computational and Theoretical Nanoscience* **6**, 625-628 (2009).
15. R. Faccio, P. A. Denis, H. Pardo, C. Goyenola, A. W. Mombru, *J. Phys.: Condens. Matter* **21**, 285304 (2009).
16. M. Topsakal, S. Ciraci, *Phys. Rev. B* **81**, 024107 (2010).
17. C. K. Gan, D. J. Srolovitz, Phys. Rev. B **81**, 125445 (2010).
18. Q. Lu, R. Huang, *Phys. Rev. B* **81**, 155410 (2010).
19. F. Liu, P. M. Ming, J. Li, *Phys. Rev. B* **76**, 064120 (2007).
20. X. Wei, B. Fragneaud, C. A. Marianetti, J. W. Kysar, *Phys. Rev. B* **80**, 205407 (2009).
21. Q. Lu, R. Huang, *Int. J. Appl. Mech.* **1**, 443-467 (2009).
22. C. Lee, X. Wei, J. W. Kysar, J. Hone, *Science* **321**, 385-388 (2008).
23. P. Koskinen, S. Malola, H. Hakkinen, *Phys. Rev. Lett.* **101**, 115502 (2008).
24. X. Jia, et al., *Science* **323**, 1701-1705 (2009).
25. M. F. Yu, O. Lourie, M. J. Dyer, K. Moloni, T. F. Kelly, R. S. Ruoff, *Science* **287**, 637-640 (2000).
26. D. W. Brenner, O. A. Shenderova, J. A. Harrison, S. J. Stuart, B. Ni, S. B. Sinnott, *J. Phys. Condens. Mat.* **14**, 783-802 (2002).
27. M. Arroyo, T. Belytschko, *Phys. Rev. B.* **69**, 115415 (2004).
28. J. Zhou, R. Huang, *J. Mech. Phys. Solids* **56**, 1609-1623 (2008).

Mater. Res. Soc. Symp. Proc. Vol. 1284 © 2011 Materials Research Society
DOI: 10.1557/opl.2011.227

Dynamics of Graphene Nanodrums

Gustavo Brunetto[1*], Sergio B. Legoas[2], Vitor R. Coluci[3], Liacir S. Lucena[4], and Douglas S. Galvao[1]

[1]Departamento de Física Aplicada, Unicamp, 13083-859 Campinas, Sao Paulo, Brazil.
[2]Departamento de Física, Universidade Federal de Roraima, 69304-000 Boa Vista, Roraima, Brazil.
[3]Faculdade de Tecnologia, Unicamp, 13484-370 Limeira, Sao Paulo, Brazil.
[4]Departamento de Física, Universidade Federal do Rio Grande do Norte, 59072-970 Natal, Brazil.
[*]Corresponding author: gusbru@ifi.unicamp.br

ABSTRACT

Recently, it was proposed that graphene sheets deposited on silicon oxide can act as impermeable atomic membranes to standard gases, such as helium, argon, and nitrogen. It is assumed that graphene membrane is clamped over the surface due only to van der Waals forces. The leakage mechanism can be experimentally addressed only indirectly. In this work we have carried out molecular dynamics simulations to study this problem. We have considered nano-containers composed of a chamber of silicon oxide filled with gas and sealed by single and multi-layer graphene membranes. The obtained results are in good qualitative agreement with the experimental data. We observed that the graphene membranes remain attached to the substrate for pressure values up to two times the largest value experimentally investigated. We did not observe any gas leakage through the membrane/substrate interface until the critical limit is reached and then a sudden membrane detachment occurs.

INTRODUCTION

From its experimental realization by Novoselov et al. [1], graphene has been the subject of intense theoretical and experimental investigations. This is in part due to its exceptional mechanical [2,3] and electronic properties [4]. Graphene has demonstrated to have nonlinear elastic properties [5,6]. These properties can be exploited to create new technological applications [1-6].

An interesting graphene application is like nano-membranes. Due to its bi-dimensional, one-atom thick material, graphene is the ultimate membrane. Membranes are very important in many physical, chemical, and biological systems [7]. Membranes are perfect structures to be used to divide the space into two regions and/or to separate systems with different physical or chemical properties.

Recently, Bunch et al. [7] proposed that graphene could be used as impermeable atomic membranes. They deposited graphene membranes over a cavity made of silicon oxide filled with different gases. The membrane was clamped on all sides only by van der Waals forces between the graphene and the cavity, creating a certain volume of confined gas (nano-container). In Figure 1 we present a schematic representation of the used systems.

They concluded that the gas leakage is mainly through the silicon oxide cavity walls and not through the graphene membranes or through graphene-oxide interface. These conclusions were obtained indirectly, once it was not possible to directly measure the gas leakage.

In this work we have carried out molecular dynamics simulations to address the problem of gas leakage and the mechanisms of membrane detachment in model systems as the one showed in Figure 1.

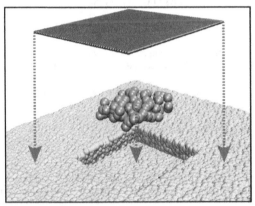

Figure 1. Scheme of a nano-container. A silicon oxide cavity (yellow) is filled with different gases (green) and sealed by graphene membranes (grey). See text for discussions.

MODELING

We carried out fully atomistic molecular dynamics (MD) simulations, in the framework of classical mechanics with standard molecular force field [8], using the parallel molecular dynamics NAMD code [9] on its CUDA implementation [10].

We have considered silicon oxide (SiO_2) cavities of different sizes, as well as, graphene membranes of different sizes and number of layers. The cavities were generated creating chambers into large SiO_2 slabs (we have considered structures containing up to six million atoms). These chambers were then filled with argon gas (one of the gases used in the experiments [7]) and sealed with graphene membranes (one up to three layers). The system is then heated up, in order to increase the gas pressure inside the chamber, and the membrane dynamics and gas leakage investigated up to the moment of membrane structural detachment. The MD simulations were carried out in the NVT ensemble, using a Langevin thermostat [9-10] with time steps of 0.5 femto seconds. Typical MD runs of up to 1 nano seconds were performed.

RESULTS AND DISCUSSION

Elasticity of the Graphene Nanomembranes

The mechanical response of the membrane due to an external applied force was investigated to determine the membrane stiffness. Two cases were considered: (i) a membrane of 20 x 20 nm^2 placed over silicon oxide cavities of 10 x 10 x 5 nm^3 and 15 x 15 x 5 nm^3; (ii) two isolated membranes (20 x 20 nm^2 and 40 x 40 nm^2) where the borders (5% of the side) are kept fixed.

The membrane, initially placed aligned to the *xy*-plane, was subjected to an external force applied at its center along the *z*-direction (Figure 2-a). A quasi- linear behavior of the displacement versus applied force was observed for the cases considered here (Figure 2-b and Figure 2-c). In addition, there is a dependence on the slope with the size of cavity (Figure 2-b). Because of the small size systems it is expected that larger holes/membranes appear to be more flexible then when compared to smaller ones. The same behavior occurs with larger membranes without substrates (Figure 2-c). These results indicated that for the structures considered here the contact region with the substrate still influences the capability to deform of the suspended membranes. However, the obtained quasi- linear behavior is similar to the one observed for macro-scale membranes [7]. This is a good indication that we are at regimes where the force field is still working properly.

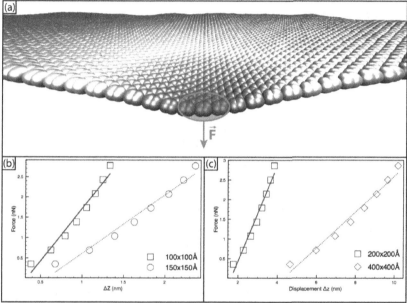

Figure 2. (a) An external force is applied at the center of the graphene membrane. Force vs displacement curves for; **(b)** a 20x20 nm^2 membrane over a cavity of different hole sizes (10 x 10x5 nm^2 and 15 x 15 x 5 nm^2), and; **(c)** two membranes with fixed borders and without supporting substrate.

Gas leakage through Silicon Oxide Walls

We have carried out simulations to determine how is the dynamical process of the argon leakage through the walls of an amorphous silicon oxide cavity. We used a hollow closed cube structure with wall thickness of 1 nm filled with argon gas (Figure 3-a). For those simulations, all the atoms of the SiO_2 are free to move, except the ones on the external edges that were fixed to keep the box shape unchanged.

During the simulations we observed that argon atoms, initially inside the box, could diffuse through the natural holes in the amorphous silicon oxide walls. Typical results are shown in Figure 3-b. This is consistent with the experimental data [7] and shows that the modeling captured the essential physics of the membrane elasticity and gas leakage.

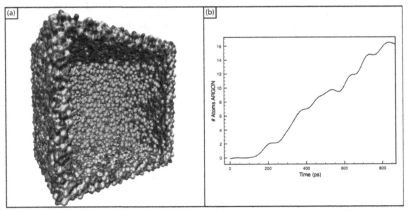

Figure 3. (**a**) Schematic representation of argon gas (in green) initially inside a silicon oxide box, with walls of 1 nm of thickness. One side of the box was made transparent to show the gas inside. (**b**) The graphic shows the number of argon atoms that leaks from inside the box during the simulation, for a gas temperature of 600 K.

Leakage through the membrane/substrate interface

Finally, we investigated the leakage process through the interface between the graphene and silicon oxide substrate (Figure 4). The 20 x 20 nm^2 graphene membrane is clamped on the surface only by van der Waals interactions. The cavity (15.4 x 15.4 x 3.1 nm^3) contained initially about 1000 argon atoms. The pressure inside the cavity was estimated using the ideal gas law.

The gas pressure/temperature is increased and the behavior of the system analyzed. We considered as reference pressure (P_0), the pressure needed to retain the membrane on its planar configuration. As the pressure starts to grow, the membrane starts to stretch. At the beginning, the stretching is not significant (Figure 4-a), but as the pressure increases (Figure 4-b) the deformations become more pronounced and the membrane bulges. As the pressure reaches a critical limit, the contact forces (mainly van der Waals ones) are overcame and they are no longer capable of keeping the membrane clamped over the substrate. At this moment one side of the membrane sudden detaches from the substrate and the gas inside the cavity escapes (Figure 4-c), followed by an immediate pressure decrease (Figure 4-f). For the case shown in Figure 4, the gas leakage begins when the gas pressure in the cavity reaches the value of $18P_0$. Larger membranes were also considered but no gas leakage or detachment was observed before the pressures reached ~$20P_0$. Until the detachment occur no gas leakage was observed between the membrane and the substrate. These results are the first theoretical validation [11], based on fully atomistic MD simulations, that indeed graphene sheets deposited on silicon oxide can act as impermeable atomic membranes to standard gases, as proposed by Bunch *et. al.* [7].

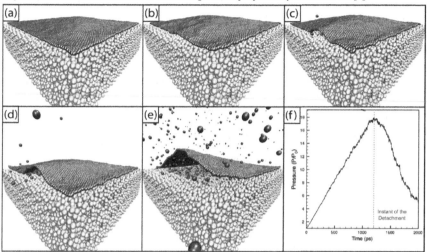

Figure 4. Snapshots from the MD simulations. (**a**) The gas pressure inside the cavity is increased, and; (**b**) the membrane starts to stretch and bulge. (**c**) As the pressure reaches a critical limit the membrane sudden detaches from the substrate and the gas inside the cavity escapes. (**f**) The calculated pressure time variation inside the cavity.

CONCLUSIONS

We have investigated, using fully atomistic molecular dynamics simulations, the problem of gas leakage in systems composed of silicon oxide nano-chambers, filled with gas and sealed by graphene membranes (one up to three layers). From our results we observed that no gas leakage occurs between the graphene membranes and the substrate until the gas pressure inside the chamber reaches a limit where the contact forces can no longer maintain the membranes clamped over the substrate. After this limit the membrane sudden detaches from the substrate and the gas escapes. These are the first theoretical validation that graphene membranes deposited over silicon substrates can indeed work as impermeable atomic membranes to standard gases, as proposed by Bunch *et. al.* [7].

ACKNOWLEDGMENTS

This work was supported by the Brazilian agencies CNPq, FINEP, FAPERN, and FAPESP. VRC acknowledges financial support of the Brazilian agencies CNPq (grants 577033/2008-5) and FAPESP (grant 2007/03923-1).

REFERENCES

1. K. S. Novoselov, A. K. Geim, S. V. Morozov, D. Jiang, Y. Zhang, S. V. Dubonos, I. V. Grigorieva, and A. A. Firsov, *Science* **306**, 666 (2004).
2. I. W. Frank, D. M. Tanenbaum, A. M. van der Zande, and P. L. McEuen, *J. Vac. Sci. Technol.* B **25**, 2558 (2007).
3. R. Faccio, P.A. Denis, H. Pardo, C. Goyenola, and A. W. Mombrú, *J. Phys.: Condens. Matt.* **21**, 285304 (2009).
4. A. K. Geim and K. S. Novoselov, *Nature Materials* **6**, 183 (2007).
5. E. Cadelano, P. L. Palla, S. Giordano, and L. Colombo, *Phys. Rev. Lett.* **102**, 235502 (2009).
6. C. Lee, X. Wei, J. W. Kysar, and J. Hone, *Science* **321**, 385 (2008).
7. J. S. Bunch, S. S. Verbridge, J. S. Alden, A. M. van der Zande, J. M. Parpia, H. G. Craighead, and P. L. McEuen, *Nano Lett.* **8**, 2458-62 (2008).
8. A. D. MacKerell, D. Bashford, M. Bellot, R. L. Dunbrack, J. Evanseck, M. J. Field, S. Fischer, J. Gao, H. Guo, S. Ha, D. Joseph, L. Kuchnir, K. Kuczera, F. T. K. Lau, C. Mattos, S. Michnick, T. Ngo, D. T. Nguyen, B. Prodhom, I. W. E. Reiher, B. Roux, M. Schlenkrich, J. Smith, R. Stote, J. Straub, M. Watanabe, J. Wiorkiewicz-Kuczera, D. Yin, and M. Karplus, J. Phys. Chem. B 102, 3586 (1998).
9. J. C. Phillips, R. Braun, W. Wang, J. Gumbart, E. Tajkhorshid, E. Villa, C. Chipot, R. D. Skeel, L. Kale, and K. Schulten, *J. Comput. Chem.* **26**, 1781 (2005). NAMD, http://www.ks.uiuc.edu/Research/namd/.
10. J. E. Stone, J. C. Phillips, P. L. Freddolino, D. J. Hardy, L. G. Trabuco, and K. Schulten, J. Comput. Chem. 28, 2618 (2007).
11. G. Brunetto, S. B. Legoas, V. R. Coluci, L. S. Lucena, and D. S. Galvao, *to be published.*

Mater. Res. Soc. Symp. Proc. Vol. 1284 © 2011 Materials Research Society
DOI: 10.1557/opl.2011.228

A Novel Method for Sorting Single Wall Carbon Nanotubes by Length

Shigekazu Ohmori[1], Takeshi Saito[1], Bikau Shukla[1], Motoo Yumura[1], Sumio Iijima[1]
[1]Nanotube Research Center, National Institute of Advanced Industrial Science and Technology, Tsukuba, Ibaraki 305-8565, Japan.

ABSTRACT

We report a novel system for sorting single wall carbon nanotubes (SWCNTs) by length via cross-flow filtration with three membrane filters of different pore sizes, 1.0, 0.45, and 0.2 μm. SWCNTs dispersed in water with the help of polymer type detergents, such as sodium carboxymethylcellulose (CMC) and polyoxyethylene stearyl ether (Brij 700), were successfully fractionated into four samples, and the atomic force microscopy (AFM) observation of those samples confirmed that their length distribution peaks are within the expected ranges from pore sizes of used filters. However, the result of the similar filtration process using a non-polymer detergent, sodium dodecylbenzenesulfonate (SDBS), showed no pronounced correlation between the length distribution of SWCNTs and the pore size. The observed difference in the sorting phenomena caused by the detergent type suggests that the permeation property depends on the complex structure resulting from the dispersed SWCNTs and detergent molecules.

INTRODUCTION

Single wall carbon nanotubes (SWCNTs) have excellent properties such as the high carrier mobility as well as the chemical stability. Thus SWCNTs are one of the most promising materials for the printed electronics and the post silicon electronic applications. It is well known that electronic properties of SWCNTs are varied from semiconducting to metallic ones depending on their structural characteristics, such as diameter, chirality, length, and so on [1–7]. Especially, the presence of metallic SWCNTs in the semiconductor channel of the transistor device arises the serious technical problem of short circuit in the channel. Needless to say, the most essential solution is the removal of metallic SWCNTs from the sample, although this technique is still under the research stage. The alternative solution is manufacturing channel by the random network of short SWCNTs [8–10]. According to ref. 10, the probability of failure in the device performance could be kept below 1 ppm by shortening the length of SWCNTs to 10 times shorter than the source-drain distance, even when the 1/3 of SWCNTs have metallic characteristics [2–4]. This strategy based on the percolation theorem suggests that controlling the length of SWCNTs is of great importance for both manufacturing electronic applications of SWCNTs as well as for understanding their fundamental properties.

As far as authors know, four techniques for sorting SWCNTs by length have been reported to date; size exclusion chromatography (SEC) [11-13], centrifugation techniques [14,15], electrophoresis [16,17], and flow field-flow fractionation [18,19]. The former two techniques possess much more excellent fractionation precision than the latter ones, although the applicable length range is limited to quite shorter range, typically less than 0.6μm. In the case of latter two techniques, in spite of their potential applicability to wider length range than SEC, they suffered from the problem of aggregating SWCNTs [16,19]. Moreover, all these previous methods generally suffer from low throughput in their batch process. Therefore, developing an efficient method for sorting SWCNTs by length is in demand for their applications. Considering the necessity of a technique having higher potentiality with greater throughput, here we propose a novel and scalable system that involves multi-steps cross-flow filtration through membrane filters of different pore sizes.

EXPERIMENT

SWCNTs used in this study were synthesized by an enhanced direct-injection pyrolytic-synthesis (eDIPS) method [20]. The purity of SWCNTs >94 atom% was characterized by the thermogravimetric analysis (TGA: Rigaku Thermo Plus Evo). The value of G/D ratio was calculated to be 173 from the resonance Raman spectrum (Jasco NRS-2100) measured by using the excitation laser line at 514.5 nm, implying the high quality of SWCNTs as well as extremely less amount of amorphous carbonaceous impurity. The diameter of SWCNTs was ranged from 1.38 to 1.68 nm confirmed by radial breathing mode (RBM) of SWCNTs [21]. The mean diameter of SWCNTs measured by the high-resolution transmission electron microscope (HR-TEM: TOPCON 002B at 100kV) was well consistent with the above characterization.

SWCNT suspensions were prepared by dispersing 4mg of SWCNTs in 200ml aqueous solutions of 1% detergent by using bath sonication. As the detergent, three different polymer- and nonpolymer-type ones, sodium carboxymethylcellulose (CMC) (MP Biomedicals low viscosity), polyoxyethylene stearyl ether (Brij 700) (Aldrich), and sodium dodecylbenzenesulfonate (SDBS) (Wako, purity > 98.5 %) were investigated for the comparison study.

The cross-flow filtration system used in the present study is a handmade apparatus converted from the commercially available filtration system, VIVAFLOW 50 (Sartorius Stedim Biotech) with an effective filtration area of 50cm^2. Hydrophilic poly tetrafluoroethylene (PTFE) membrane (Millipore Omnipore) was applied as the filter. As shown in Figure 1a, SWCNT suspension was introduced into the filter system at the flow rate of 1ml/sec, and the retentate was continuously returned to the source suspension by the peristaltic pump with the Tigon tubing, while the permeated suspension was flowed out of the system and collected. The filtration continued until stopping the permeation. Note that the stop of the permeation was caused not by the clogging of SWCNTs on the filter membrane, but by the osmotic pressure induced by the increasing concentration of dispersants during the filtration. The retentate was stored for further characterization. On the other hand, the permeate was again processed through the further similar filtration. We repeated the filtration processes by using the membrane filter of 1.0, 0.45, and 0.2μm pore sizes, stepwise in decreasing order of the pore size. Consequently SWCNTs suspension is, at most, separated to four samples, which were three retentates and one last permeate by the three steps filtration.

Separated samples were characterized by optical absorption spectroscopy (Hitachi, U-4100) and atomic force microscopy (AFM: SII NanoTechnology Inc., SPA400) observations for estimating the relative amount against control and the length distribution of SWCNTs. Specimens for AFM measurements were prepared by following procedures. Silicon (Si) substrates were chemically functionalized by using aminopropyltrimethoxysilane (APTES). SWCNT suspensions were dropped on the functionalized Si substrates and washed with water to remove the detergent. This procedure was repeated until the frequency of SWCNTs in the AFM image became suitable for investigating the distribution of their length. Lengths of more than 300 SWCNTs were measured in each sample for the statistical analysis.

DISCUSSION

Four samples A–D, that is, three retentates of filtrations using membranes with pore sizes of 1.0, 0.45, and 0.2 μm and one permeate of the last step resulted from fractionating SWCNTs dispersed in the CMC solution as shown in Figure 1b. Figure 1b also shows the photographic images of A–D. The difference in the turbidity of these samples shows the greater amounts of SWCNTs in A and B than those in C and D.

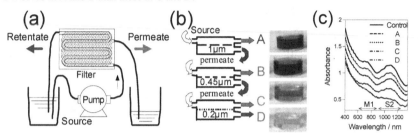

Figure 1. A schematic depiction of the cross-flow filtration system (a), the three-steps fractionation processes and sorted samples A–D (b), and their optical absorption spectra with that of the source suspension as a control (c).

Optical absorption spectra of those samples measured in the wavelength range of 400–1350 nm are plotted in Figure 1c for the qualitative and quantitative analyses to estimate the filtration efficiency. Profiles of C and D are magnified 2.5 and 7 times because of their lower concentrations. For comparison, the absorption spectrum of the source suspension is also shown as a control by the four times reduced intensity. Absorption peaks originated from the second interband transition of semiconducting SWCNTs (S2) and the first interband transition of metallic SWCNTs (M1) [22] are observed in all spectra in which peak positions and shapes of absorption spectra of A–C are approximately the same with that of the control in whole measured wavelength range. In the case of D, peak positions of S2 and M1 seem to be similar to others although the spectrum feature is less resolved due to the low contents of SWCNTs. These peak positions are consistent with the value of mean diameter in used SWCNTs [22]. This qualitative comparison supports the negligible effect of the fractionation process on the compositions of SWCNTs in terms of not only the diameter distribution and the metal/semiconductor ratio, but also the featureless absorbers in the background parts, except for the length distribution.

On the basis of the above discussion, the relative amount of SWCNTs in each sample against the control can be estimated by comparing measured absorbances in the whole measurement range. By using values of the relative amount, the fractionation efficiency with one standard deviation in each step was calculated as follows: A: 29±0.5 %, B: 33±0.3 %, C: 10±0.2 %, and D: 2.9±0.5 %. Narrow standard deviations show the stability of dispersed SWCNTs composition in these samples. The total amount of SWCNTs fractionated by these processes, that is the summation of all the fractionation efficiencies, suggests the loss of 25% SWCNTs in the source suspension through these processes, probably due to the aggregation of SWCNTs on the membranes and the inside of the system.

Figure 2a shows typical AFM topographic images of separated samples A–D. From the analysis of these images, the heights of individual SWCNTs less than 3nm indicate dispersing SWCNTs into debundled or few-bundled ones.

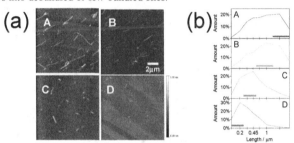

Figure 2. Typical topographic AFM images (a) and length distribution profiles (b) of SWCNTs in samples A–D.

Length distribution profiles of A–D from AFM measurements are plotted in Figure 2, in which these distribution data are expressed as the percentage of measured counts in each sample with the horizontal axis scaled as log10. Peak positions of the length distributions of A–D are consistent with the ranges expected from pore sizes of used filters, that clearly shows the length fractionation of SWCNTs by the multi-steps cross-flow filtration. Particularly in B–D, more than 41% of SWCNTs are in the expected range supporting the greater potentiality of the fractionation process. Sharpening in length distribution observed in C and D is reasonable, because the time taken for the fractionation in these samples is consequently longer.

In order to investigate the impact of the detergent type, the similar experimental procedures of fractionation were carried out for SWCNT suspensions of SDBS and Brij 700 detergents, that have been frequently used for dispersing SWCNTs. Surprisingly, the SWCNTs/SDBS suspension permeated through all filters with the considerable amount of SWCNT aggregates on filter membranes. The last permeate contained only 7 % of SWCNTs against the control on the basis of the optical absorption analysis. Furthermore, the AFM observation shows no correlation between their length distribution profile and the pore size of used membrane.

The main cause of the abovementioned different filtration behavior in SWCNTs/SDBS suspension is probably due to the removal of SDBS molecules from SWCNTs' surface by the turbulence in the filtration system. This result is quite similar to the previous work on the cross-flow filtration of multi-wall carbon nanotubes (MWCNTs) suspension prepared by using sodium dodecyl sulfate, showing low or negative correlation between the average lengths of MWCNTs and pore sizes of filter membranes [23]. We also investigated the filtration of SWCNT suspension using Brij 700 solution and confirmed the result similar to the case of CMC. Because Brij700 is composed by the long polymer chain as a hydrophilic part, and the hydrocarbon chain as a hydrophobic part, that is the mixed characteristics of CMC and SDBS, the observed difference in the present filtration results between SDBS and polymer-type detergents might be due to the lower aspect ratio of micelles in polymer detergents as illustrated in Figure 3.

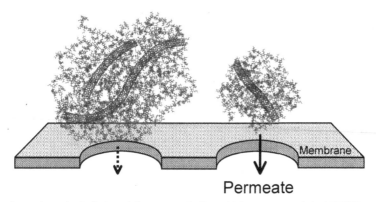

Permeate

Figure 3. A schematic depiction of dispersant micelles with low aspect ratio in SWCNTs suspensions dispersed by polymer-type detergents.

The lower aspect ratio of CMC micelles is expected from the long polymer chain (degree of polymerization: ~400) with the high molecular weight (~90 kDa). The micelle size of Brij 700 was reported to be around 20 nm by the small angle X-ray scattering measurement (SAXS) [24]. Therefore, CMC and Brij 700 can wrap SWCNTs and embed them in the matrix of micelle form, which decreases the aspect ratio of their micelles enabling them to be sorted by the size. In other words, the permeation property of the dispersants, consisting of SWCNTs and detergent molecules, depends on their resulting complex structure. In addition, such complex structure also causes the difference in the aggregation property of SWCNTs observed in the present study. Thus, it can be concluded that the key of this fractionation system is the employment of polymer type detergents.

CONCLUSIONS

A novel length fractionation system of SWCNTs with the multi-steps cross-flow filtration technique using three membrane filters of different pore sizes, 1.0μm, 0.45μm, and 0.2μm, has been developed that successfully separated a suspension of SWCNTs dispersed with the help of polymer detergents, CMC and Brij700, into four resulting samples; three retentates and one permeate of the last step. The analysis of their optical absorption spectra showed that 75% of the starting SWCNTs were fractionated leaving 25% probably on the filters and inner walls of the tubing as aggregates. AFM observations confirmed that the length distribution peaks in profiles of four separated samples are within the expected ranges from pore sizes of used filters. Additionally, it was found that both the permeation and re-aggregation properties of dispersed SWCNTs in the filtration process greatly depend on their complex structure with detergent molecules.

ACKNOWLEDGMENTS

This work is partially supported by the New Energy and Industrial Technology Development Organization (NEDO).

REFERENCES

1. K.Tanaka, K. Okahara, M. Okada, Y. Yamabe, *Chem. Phys. Lett.*, **191**, 469–472 (1992).
2. R. Saito, M. Fujita, G. Dresselhaus, M. S. Dresselhaus, *Phys. Rev. B* **46**, 1804–1811 (1992).
3. N. Hamada, S. Sawada, A. Oshiyama, *Phys. Rev. Lett.* **68**, 1579–1581 (1992).
4. J. A. Fagan, J. R. Simpson, B. J. Bauer, S. H. De Paoli Lacerda, M. L. Becker, J. Chun, K. B. Migler, A. R. Hight Walker, E. K. Hobbie, *J. Am. Chem. Soc.* **129**, 10607–10612 (2007).
5. A. Rajan, M. S. Strano, D. A. Heller, T. Hertel, K. Schulten, *J. Phys. Chem. B* **112**, 6211–6213 (2008).
6. X. Sun, S. Zaric, D. Daranciang, K. Welsher, Y. Lu, X. Li, H. Dai, *J. Am. Chem. Soc.* **130**, 6551–6555 (2008).
7. T. Nakanishi, T. Ando, *J. Phys Soc. Jpn.* **78**, 114708 (2009).
8. G. E. Pike, C. H. Seager, *Phys. Rev. B* **10**, 1421–1434 (1974).
9. M. Ishida, F. Nihey, *Appl. Phys Lett.* **92**, 163507 (2008).
10. Y. Asada, Y. Miyata, K. Shiozawa, Y. Ohno, R. Kitaura, T. Mizutani, H. Shinohara, *J. Phys. Chem. C* (in press).
11. G. S. Duesberg, J. Muster, V. Krstic, M. Burghard, S. Roth, *Appl. Phys. A* **67**, 117–119 (1998).
12. M. Zheng, A. Jagota, E. D. Semke, B. A. Diner, ROBERT S. Mclean R. S., S. R. Lustig, R. E. Richardson, N. G. Tassi, *Nat. Mater.* **2**, 338–342 (2003).
13. X. Huang, R. S. Mclean, M. Zheng, *Anal. Chem.* **77**, 6225–6228 (2005).
14. J. A. Fagan, M. L. Becker, J. Chun, P. Nie, B. J. Bauer, J. R. Simpson, A. Hight-Walker, E. K. Hobbie, *Langmuir* **24**, 13880–13889 (2008).
15. J. A. Fagan, M. L. Becker, J. Chun, E. K. Hobbie, *Adv. Mater.* **20**, 1609–1613 (2008).
16. S. K. Doorn, R. E. Fields, H. Hu, M. A. Hamon, R. C. Haddon, J. P. Selegue, V.Majidi, *J. Am. Chem. Soc.* **124**, 3169–3174 (2002).
17. A. Daniel, D. A. Heller, R. M. Mayrhofer, S. Baik, Y. V. Grinkova, M. L. Usrey, M. S. Strano, *J. Am. Chem. Soc.* **126**, 14567-14573 (2004).
18. B. Chen, J. P. Selegue, *Anal. Chem.* **74**, 4774–4780 (2002).
19. J. Chun, J. A. Fagan, E. K. Hobbie, B. J. Bauer, *Anal. Chem.* **80**, 2514–2523 (2008).
20. T. Saito, S. Ohshima, T. Okazaki, S. Ohmori, M. Yumura, S. Iijima, *J. Nanosci. Nanotechnol.* **8**, 6153–6157 (2008).
21. A. Jorio, R. Saito, J. H. Hafner, C. M. Lieber, M. Hunter, T. McClure, G. Dresselhaus, M. S. Dresselhaus, *Phys. Rev. Lett.* **86**, 1118–1121 (2001).
22. T. Saito, S. Ohmori, B. Shukla, M. Yumura, S. Iijima, *Appl. Phys. Exp.* **2**, 095006 (2009).
23. T. Abatemarco, J. Stickel, J. Belfort, B. P. Frank, P. M. Ajayan, G. Belfort, *J. Phys. Chem. B* **103**, 3534-3538 (1999).
24. C. Sommer and J. S. Pedersen, *Langmuir* **21**, 2137–2149 (2005).

AUTHOR INDEX

SUBJECT INDEX

Printed in the United States
By Bookmasters